Mathematical Research

Nonlinear Dynamics and Quantum Dynamical Systems

edited by

G. A. Leonov
V. Reitmann
W. Timmermann

Volume 59

AKADEMIE-VERLAG BERLIN

In this series original contributions of mathematical research in all fields are contained, such as

- research monographs
- collections of papers to a single topic
- reports on congresses of exceptional interest for mathematical research.

This series is aimed at promoting quick information and communication between mathematicians of the various special branches.

In diese Reihe werden Originalbeiträge zu allen Gebieten der mathematischen Forschung aufgenommen wie

- Forschungsmonographien
- Sammlungen von Arbeiten zu einem speziellen Thema
- Berichte von Tagungen, die für die mathematische Forschung besonders aktuell sind.

Die Reihe soll die schnelle Information und gute Kommunikation zwischen den Mathematikern der verschiedenen Fachgebiete fördern.

Manuscripts
in English and German comprising at least 100 pages and not more than 500 pages can be admitted to this series. With respect to a quick publication the manuscripts are reproduced photomechanically. Authors who are interested in this series please turn directly to the 'Akademie-Verlag'. Here you will get more detailed information about the form of the manuscripts and the modalities of publication.

Manuskripte
in englischer und deutscher Sprache, die mindestens 100 Seiten und nicht mehr als 500 Seiten umfassen, können in diese Reihe aufgenommen werden. Im Interesse einer schnellen Publikation werden die Manuskripte auf fotomechanischem Weg reproduziert. Autoren, die an der Veröffentlichung entsprechender Arbeiten in dieser Reihe interessiert sind, wenden sich bitte direkt an den Akademie-Verlag. Sie erhalten dort genauere Informationen über die Gestaltung der Manuskripte und die Modalitäten der Veröffentlichung.

Nonlinear Dynamics and
Quantum Dynamical Systems

Mathematical Research - Mathematische Forschung

Wissenschaftliche Beiträge
herausgegeben von der
Akademie der Wissenschaften der DDR
Karl-Weierstraß-Institut für Mathematik

Band 59

Nonlinear Dynamics and
Quantum Dynamical Systems

Nonlinear Dynamics and Quantum Dynamical Systems

Contributions to the International Seminar ISAM–90 held in Gaussig (GDR), March 19–23, 1990

edited by

Gennadij A. Leonov
Volker Reitmann
Werner Timmermann

Akademie-Verlag Berlin 1990

Herausgeber:

Prof. Dr. Gennadij A. Leonov
Leningrad State University
Department of Mathematics and Mechanics

Doz. Dr. Volker Reitmann
Prof. Dr. Werner Timmermann
Technische Universität Dresden
Sektion Mathematik

Die Titel dieser Schriftenreihe werden vom Originalmanuskript der Autoren reproduziert.

ISBN 3-05-500858-8
ISSN 0138-3019

Erschienen im Akademie-Verlag Berlin,Leipziger Str. 3-4,Berlin,DDR-1086

Lizenznummer: 202.100/404/90
Printed in the German Democratic Republic
Gesamtherstellung: VEB Kongreß- und Werbedruck,Oberlungwitz,DDR-9273
Lektor: Dr. Reinhard Höppner
LSV 1065
Bestellnummer: 764 164 9 (2182/59)

02200

P R E F A C E

This volume compiled in preparation of the workshop ISAM 90 collects survey and original articles on nonlinear dynamics and quantum dynamical systems.
These topics are assumed to be of interest for a large number of scientists.
In the centre there are contributions to the followings problems: estimation of Hausdorff dimension of strange attractors of the Lorenz and Duffing equations, dynamical systems on Riemannian manifolds and bifurcations; entropy, positivity and automorphisms in operator algebras and their relation with quantum dynamics.
The editors are indebted to all the authors who contributed to these proceedings. They sincerely thank Dr.R.Höppner of the Akademie-Verlag, Berlin for his kind expressions of interest and encouragement.

G.A.Leonov
Leningrad

V.Reitmann
W.Timmermann
Dresden

January 1990

C O N T E N T S

Linear Dynamic Processes with Inhomogeneous Time Scale

Bernd Aulbach[1)2)]
Stefan Hilger[1)2)]

Introduction

In the mathematical description of dynamic processes the time is commonly played either by a real or an integer variable. Which one is appropriate in a particular context depends both on the modelled phenomenon and the mathematical approach. On one hand the rich and well developed theory of differential equations provides a broad mathematical basis for an understanding of continuous time dynamic processes. On the other hand the classical theory of difference equations is less helpful because of its lack of qualitative features. This is particularly problematic since the discrete time viewpoint experiences an enormous renaissance both from the point of view of discretization of continuous time systems and the direct description of discrete time models. There are many results on continuous time dynamical systems which are needed in a discrete time context. In such a case those results - not the proofs - are usually translated into the discrete time language and then they are applied without any caution. This, however, is dangerous. It should be known that in some prominent cases things go wrong. After all, continuous time orbits and discrete time orbits are topologically quite different. On the other hand, should any result on dynmaical systems be proved twice in spite of the well known duality?

It is our opinion that it is best to have a mathematical rigorous way to consider both kinds of systems at the same time. In this paper we want to present a glance at a certain kind of calculus which comprises those features of the differential and difference calculus as they are relevant for the development of a qualitative theory of dynamical systems. Because of the lack of space we confine our considerations in this note to linear systems. For a more general treatment we refer to Hilger [3] and Aulbach and Hilger [1].

In this paper by a time scale we mean any closed subset of $\mathbb{R}$, the set of reals . Thus, the common "homogeneous" time scales $\mathbb{R}, \mathbb{N}, \mathbb{Z}, h\mathbb{Z}, h > 0$ (constant step size) are contained as special cases. We, however, allow any closed subset of $\mathbb{R}$ to be a time scale, that's why we call our times scales "inhomogeneous". It will turn out that the results we are going to present are equally valid for homogeneous and inhomogeneous time scales. That means, our apprach does not only represent a unification of certain results on continuous time and discrete time dynamical systems but it also provides a new direction of research: dynamical processes with inhomogenous time scale.

1) Institut für Mathematik, Universität Augsburg, Universitätsstr. 8, D-8900 Augsburg, FRG
2) Work supported by Deutsche Forschungsgemeinschaft, Au 66/3-1

This paper is devided in two parts. In part one we briefly describe some features of time scales, functions defined on time scales and operations with those functions. Part two is concerned with some features of the qualitative theory of linear dynamic processes. For lack of space we will give only one proof in full detail. Nevertheless, we wish to mention that most proofs are modelled after the continuous time case. In particular, the induction principle we are going to present in this note allows to mimic some proofs from Dieudonné [2]. In any case, detailed proofs can be found in Hilger [3].

Part I: Calculus on time scales

1. Time scales

As mentioned above by a time scale we mean any closed subset of $\mathbb{R}$. Time scales are described by the symbol $\mathbf{T}$. Examples of time scales are $\mathbb{N}, \mathbb{Z}, \mathbb{R}, \{x \in \mathbb{R} : x \leq 0\} \cup \{\frac{1}{n} : n \in \mathbb{N}\}$ or the Cantor set. $\mathbb{Q}$ and $\{x \in \mathbb{R}\backslash\mathbb{Q} : 0 \leq x \leq 1\}$ are no time scales.

As subsets of $\mathbb{R}$ time scales carry an order structure as well as a topological structure in a canonical way. There is, however, in general no algebraic structure on $\mathbf{T}$. We first describe now some consequences of the embedding of $\mathbf{T}$ in $\mathbb{R}$ and the closedness of $\mathbf{T}$.

A. Order structure: A time scale $\mathbf{T}$ may be bounded above or below. All order theoretical notions such as bounds, least upper bounds, greatest lower bounds and intervals are available in $\mathbf{T}$ as they are in $\mathbb{R}$. We particularly want to mention that any bounded subset A of $\mathbf{T}$ has a least upper bound and a greatest lower bound, both belonging to $\mathbf{T}$, not necessarily to A. By an interval we always mean in this text the intersection of a real interval with a given time scale.

B. Topological structure: Because of the closedness of $\mathbf{T}$ the notion of closedness in $\mathbf{T}$ coincides with closedness in $\mathbb{R}$. In view of the previous order theoretical statements the same is true for compactness. A problem, however, arises with the notion of openness. Obviously any subset A of $\mathbf{T}$ which is open in $\mathbb{R}$ is also open in $\mathbf{T}$. The reverse is in general not true, though, as the simple example $\mathbf{T} := \mathbb{Z}$ shows, where any subset is (in the induced topology) open in $\mathbf{T}$ but not open in $\mathbb{R}$. We will take care of this by distinguishing between $\mathbb{R}$-opennes and $\mathbf{T}$-openness. A neighborhood in this text is always a real neighborhood intersected with $\mathbb{R}$.

C. Jump a operators: Obviously a time scale $\mathbf{T}$ may be connected or not. In order to overcome this topological deficiency we introduce the concept of jump operators. The mappings

$$\sigma : \begin{cases} \mathbf{T} \to \mathbf{T} \\ t \mapsto inf\{s \in \mathbf{T} : s > t\} \end{cases}, \quad \rho : \begin{cases} \mathbf{T} \to \mathbf{T} \\ t \mapsto sup\{s \in \mathbf{T} : s < t\} \end{cases}$$

are called jump operators. In case $\mathbf{T}$ is bounded above we supplement the definition by σ (max $\mathbf{T}$) := max$\mathbf{T}$ and acordingly ρ(min$\mathbf{T}$):= min$\mathbf{T}$ if $\mathbf{T}$ is bounded below.

The jump operators allow us to classify the points of a time scale $\mathbf{T}$. A point $t \in \mathbf{T}$ is called

$$\begin{aligned}&\text{right-dense, if } \sigma(t) = t\\ &\text{right-scattered, if } \sigma(t) > t,\\ &\text{left-dense, if } \rho(t) = t,\\ &\text{left-scattered, if } \rho(t) < t.\end{aligned}$$

In view of the fact that the difference calculus to be included in our calculus is not symmetric in time we break up the time symmetry at this point by preferring the positive i.e. increasing time direction. We define the so-called grainyness

$$\mu^* : \begin{cases} \mathbf{T} \to \mathbb{R}_0^+ \\ t \mapsto \sigma(t) - t. \end{cases}$$

Obviously the grainyness is identically 0 in case $\mathbf{T} := \mathbb{R}$ and identically 1 in case $\mathbf{T} := \mathbb{Z}$.

If a time scale $\mathbf{T}$ has a maximal element which moreover is left-scattered, then this point plays a particualar role in several respects and therefore we call it degenerate. All other elements of $\mathbf{T}$ are called non-degenerate and the subset of non-degenerate points of $\mathbf{T}$ is denoted by $\mathbf{T}^\kappa$.

2. <u>Induction principle</u>

Although we hardly can give detailed proofs in this short note we wish to mention a basic tool which is used in many proofs. It is a principle using an induction argument just suited for time scales.

<u>Theorem 1</u>: Let $\mathbf{T}$ be a time scale with minimal element t_0. Suppose for any $t \in \mathbf{T}$ there is a statement $A(t)$ such that the following conditions are satisfied:

(I) $A(t_0)$ is true,

(II) If t is right-scattered and $A(t)$ is true, then also $A(\sigma(t))$ is true,

(III) For each right-dense t there exists a neighborhood U with the following property: if $A(t)$ is true, then $A(s)$ is true for all $s \in U$ with $s \geq t$,

(IV) If t is left-dense and $A(s)$ is true for all $s \in [t_0, t)$, then $A(t)$ is true.

Then $A(t)$ is true for all $t \in \mathbf{T}$.

<u>Remark</u>: In case $\mathbf{T} := \mathbb{N}$ the conditions (III) and (IV) disappear and Theorem 1 becomes the well known principle of mathematical induction.

The next step in the development of a calculus on time scales towards a theory of dynamic processes is to consider functions which are defined on a time scale $\mathbf{T}$ and take their values in an appropriate space X. If X is a topological space the concept of continuity arises in a straightforward manner due to the embedding of $\mathbf{T}$ in $\mathbb{R}$. Therefore we do not persue continuity here any further. For the concept of linear approximation, i.e. differentiation, however, the topological structure of $\mathbf{T}$

plays a vital role. In fact the generally lacking openness of $\mathbf{T}$ requires a particular proceeding when one aims at a concept of differentiability which contains as special cases the differential calculus on one hand and the difference calculus on the other.

3. Differentiation

A mapping $f : \mathbf{T} \to X$ from a time scale $\mathbf{T}$ to a real Banach space X is called differentiable at $t_0 \in \mathbf{T}$, if there exists an $a \in X$ with the following property: for any $\varepsilon > 0$ there exists a neighborhood U of t_o, such that

$$|f(\sigma(t_0)) - f(t) - (\sigma(t_0) - t)a| \leq \varepsilon|\sigma(t_0) - t| \text{ for all } t \in U.$$

$f : \mathbf{T} \to X$ is called differentiable, if f is differentiable at any $t \in \mathbf{T}$.

Remark: It is easily seen that in the two special cases $\mathbb{R}$ and $\mathbb{Z}$ the "derivative" a is uniquely determined, in fact one gets $a = \frac{df}{dt}(t_o)$ and $a = f(t_0 + 1) - f(t_0)$, respectively.

That also in the general case the "derivative" is uniquely determind at any non-degenerate point in $\mathbf{T}$ is the content of

Theorem 2: (i) For any $t_0 \in \mathbf{T}^\kappa$ the function f has at most one $a \in X$ satisfying the definition of differentiability. In case such an $a \in X$ exists it is denoted by $f^\Delta(t_0)$ and called derivative of f at t_0.
(ii) If $\mathbf{T}$ has a left-scattered maximum t^*, then at t^* any function $f : \mathbf{T} \to X$ is differentiable and the definition of differentiability can be satisfied with any $a \in X$.

Remark: This theorem demonstrates the extraordinary role of a left-scattered maximum, if it exists.

The relation between continuity and differentiability is described in

Theorem 3: For a given function $f : \mathbf{T} \to X$ and $t_0 \in \mathbf{T}^\kappa$ the following is true:
(i) If f is differentiable at t_0, then f is continuous at t_0.
(ii) If f is continuous at t_0 and t_0 is right-scattered, then f is differentiable at t_0 with derivative $f^\Delta(t_0) = \frac{f(\sigma(t_0)) - f(t_0)}{\mu^*(t_0)}$.

The linearity of the differentiation is obvious and needs no further mention. The product rule, however, deserves an explicit formulation.

Theorem 4: (i) Let X_1, X_2, X_3 be Banach spaces and $\bullet \begin{cases} X_1 \times X_2 \to X_3 \\ (x_1, x_2) \mapsto x_1 \cdot x_2 \end{cases}$ be a bilinear continuous mapping. Then, if $f : \mathbf{T} \to X_1$ and $g : \mathbf{T} \to X_2$ are differentiable at $t_0 \in \mathbf{T}^\kappa$, then also $f \cdot g : \mathbf{T} \to X_3$ is differentiable at t_0 with derivative

$$(f \cdot g)^\Delta(t_0) = f(\sigma(t_0)) \cdot g^\Delta(t_0) + f^\Delta(t_0) \cdot g(t_0).$$

(ii) Let X be a Banach algebra with unity I and suppose $f, g : \mathbf{T} \to X$ satisfy $f(t) \cdot g(t) \equiv I$.
Then, if f is differentiable at $t_0 \in \mathbf{T}^\kappa$, so is g and its derivative is

$$g^\Delta(t_0) = -g(\sigma(t_0)) \cdot f^\Delta(t_0) \cdot g(t_0).$$

The mean value theorem in our calculus on time scales appears in the following form:

Theorem 5: Let $t_1, t_2 \in \mathbf{T}, t_1 < t_2$ be given and suppose $f : \mathbf{T} \to X$ and $g : \mathbf{T} \to \mathbb{R}$ are differentiable. Then if

$$|f^{\Delta}(t)| \leq g^{\Delta}(t) \text{ holds true for all } \quad t \in [t_1, t_2]^{\kappa},$$

then one gets the estimate

$$|f(t_2) - f(t_1)| \leq g(t_2) - g(t_1).$$

Remarks: 1. The proof of this theorem (cf. [3]) is typical for our calculus on time scales. One uses a certain "ε-margin" (here the term $\varepsilon(t - t_1)$) and proves by means of the induction principle, that the statement

$$A(t) : |f(t_2) - f(t_1)| \leq g(t_2) - g(t_1) + \varepsilon(t - t_1)$$

is true for all t. Then the limit $\varepsilon \to 0$ does the job. An explicit example of this kind of proof will be given at the end of this paper (proof of Theorem 12).

2. It is an immediate consequence of this theorem that a function with everywhere vanishing derivative is constant. This allows us to intruduce the notion of an intergral through the concept of antidifferentiation.

4. Antidifferentiation

For a given function $g : \mathbf{T}^{\kappa} \to X$ and a subset A of $\mathbf{T}$ a function $f : A \to X$ is called an antiderivative of g on A, if f is differentiable on A such that

$$f^{\Delta}(t) = g(t) \text{ for all } t \in A^{\kappa}.$$

In order to obtain an appropriate class of functions having antiderivatives we define the notion of rd-continuity. A function $g : \mathbf{T}^{\kappa} \to X$ is rd-continuous, if it is continuous in every right-dense point $t \in \mathbf{T}^{\kappa}$ and if the limit $\lim_{s \to t-0} g(s)$ exists for each left-dense $t \in \mathbf{T}^{\kappa}$. The class of rd-continuous functions turns out to be a "natural" class within the context of our calculus. The function $\mu^* : \mathbf{T} \to \mathbb{R}$ in case $\mathbf{T} := [0, 1] \cup \mathbb{N}$ e.g. is rd-continuous but not continuous in 1.

The basic result for the existence of antiderivatives is

Theorem 6: Any rd-continuous function $g : \mathbf{T}^{\kappa} \to X$ has an antiderivative.

Recalling the fact that a differentiable function with vanishing derivative is constant it is possible by means of Theorem 6 to define the Cauchy-integral for rd-continuous functions. Let $g : \mathbf{T}^{\kappa} \to X$ have an antiderivative $f : \mathbf{T}^{\kappa} \to X$. Then for any $r, s \in \mathbf{T}^{\kappa}, r \leq s$ we define

$$\int_r^s g(t) \Delta t := f(s) - f(r).$$

The integral is obviously linear. Other properties are stated in

Theorem 7: (i) In the framework of Theorem 4, (i) with rd-continuous functions f and g we have the formula for integration by parts

$$\int_r^s f(\sigma(t))g^{\Delta}(t)\Delta t + \int_r^s f^{\Delta}(t)g(t)\Delta t = f(s)g(s) - f(r)g(r).$$

(ii) Suppose $f : \mathbf{T}^{\kappa} \to X, |f| : \mathbf{T}^{\kappa} \to X, g : \mathbf{T}^{\kappa} \to \mathbb{R}_0^+$ have antiderivatives. Then the inequality

$$|f(t)| \leq g(t) \text{ on } [r,s]^{\kappa}$$

implies

$$|\int_r^s f(t)\Delta t| \leq \int_r^s |f(t)|\Delta t \leq \int_r^s g(t)\Delta t \leq \sup_{t\in[r,s]^{\kappa}} g(t)(s-r).$$

Part II: Linear dynamic equations

After having sketched essential parts of the calculus on times scales we are now in a position to define equations describing dynamic processes.

5. Existence and uniqueness of solutions

First of all we notice that in contrast to differential equations the solutions of difference equations are in general not uniquely determind in backward time. This means we have to find a condition which guarantees unique backward continuation of solutions.

Let $\mathbf{T}$ be a time scale and X be a real Banach space. Furthermore denote by $(\mathcal{B}(X), |\cdot|)$ the Banach algebra with unity of linear continuous operators from X to X. Then a mapping

$$A : \mathbf{T}^{\kappa} \to \mathcal{B}(X) \text{ is called } \underline{\text{regressive}},$$

if for each $t \in \mathbf{T}^{\kappa}$ the mapping

$$A(t)\mu^*(t) + id : X \to X$$

is invertible. This is the case e.g. if

$$|A(t)\mu^*(t)| < 1 \quad \text{for all } t \in \mathbf{T}$$

(von Neumann-series). Obviously in case $\mathbf{T} = \mathbb{R}$ any A is regressive (since $\mu^* = 0$) and in case $\mathbf{T} = \mathbb{Z}$ the set A is regressive e.g. if $|A(t)| < 1$ (since $\mu^* \equiv 1$).

Definition: Suppose $A : \mathbf{T}^{\kappa} \to \mathcal{B}(X)$ is rd-continuous and regressive and $b : \mathbf{T}^{\kappa} \to X$ is rd-continuous. Then a mapping $x : \mathbf{T} \to X$ is called a solution of the linear dynamic equation

$$x^{\Delta} = A(t)x + b(t), \tag{5.1}$$

(on $\mathbf{T}$), if $x^{\Delta}(t) = A(t)x(t) + b(t)$ for all $t \in \mathbf{T}^{\kappa}$. Beside the inhomogeneous (if $b \not\equiv 0$) equation (5.1) we also consider the associated homogeneous equation

$$x^{\Delta} = A(t)x. \tag{5.2}$$

If a solution $x(\cdot)$ of (5.1) in addition satisfies the condition $x(\tau) = \xi$ for a pair $(\tau, \xi) \in \mathbf{T} \times X$, it is called a solution of the initial value problem

$$x^{\Delta} = A(t)x + b(t), x(\tau) = \xi. \tag{5.3}$$

After these preparations we are ready to formulate the global existence and uniqueness theorem.

Theorem 8: (i) The initial value problem

$$x^{\Delta} = A(t)x + b(t), x(\tau) = \xi \tag{5.4}$$

has exactly one solution on $\mathbf{T}$, if $A : \mathbf{T}^{\kappa} \to \mathcal{B}(X)$ is rd-continuous and regressive and $b : \mathbf{T}^{\kappa} \to X$ is rd-continuous.
(ii)Consider the initial value problem

$$x^{\Delta} = A(t)x, \quad x(\tau) = I \tag{5.5}$$

in the Banach algebra $\mathcal{B}(X), I$ being its unity. Then if $\Phi_A(\cdot, \tau)$ denotes its solution, the solution of (5.4) has the form

$$x(t) = \Phi_A(t, \tau)\xi + \int_{\tau}^{t} \Phi_A(t, \sigma(s))b(s)\Delta s \tag{5.6}$$

(Variation of constants formula).

Remark: Notice that without the assumption of regressivity this theorem becomes false. The initial value problem

$$x^{\Delta} = -x, \quad x(2) = \xi$$

in $\mathbb{R}$ over the time scale $\mathbb{N}$ (in classical notation it is $x_{t+1} = 0$) has infinitely many solutions for $\xi = 0$ and none for $\xi \neq 0$. The regressivity condition is violated since $A(t)\mu^*(t) + id$ in this case is the zero-mapping.

6. Transition operators

By means of Theorem 8 there exists a unique solution $\Phi_A(\cdot, \tau) : \mathbf{T} \to \mathcal{B}(X)$ of the initial value problem

$$x^{\Delta} = A(t)x, \; x(\tau) = I, \tag{6.1}$$

it is called fundamental solution. The linear operator $\Phi_A(t, \tau) \in \mathcal{B}(X)$ is then called the transition operator corresponding to (6.1).

In order to formulate the expected properties of transition operators we need an algebraic concept. The set

$\mathrm{CrdR}(\mathbf{T}^\kappa, \mathcal{B}(X)) :=$
$\{A : \mathbf{T}^\kappa \to \mathcal{B}(x) | A$ rd-continuous and $A(t)\mu^*(t) + I$ invertible for each $t \in \mathbf{T}^\kappa\}$

becomes a group, the so-called regressive group, when we define the operation

$$(A\tilde{+}B)(t) := A(t)B(t)\mu^*(t) + A(t) + B(t)$$

on CrdR $(\mathbf{T}^\kappa, \mathcal{B}(X))$. We notice that the inverse for this operation is

$$(\tilde{-}A)(t) := A(t)[A(t)\mu^*(t) + I]^{-1}$$

and that the operation $\tilde{+}$ is commutative, if the product in $\mathcal{L}(X)$ is commutative.

Theorem 9: Suppose $A, B \in CrdR(\mathbf{T}^\kappa, \mathcal{B}(X))$ and $r, s, t, \tau \in \mathbf{T}$. Then we get:

(i) $\Phi_A(t,s)\Phi_A(s,r) = \Phi_A(t,r)$, in particular $\Phi_A(t,s)^{-1} = \Phi_A(s,t)$.

(ii) If A is independent of t, then
$A\int_r^s \Phi_A(t,\tau)\Delta t = \Phi_A(s,\tau) - \Phi_A(r,\tau)$.

(iii) If $\mathcal{B}(X)$ is commutative, then
$\Phi_{A\tilde{+}B}(t,s) = \Phi_A(t,s)\Phi_B(t,s)$ and $\Phi_{\tilde{-}A}(t,s) = \Phi_A(t,s)^{-1}$.

7. The logarithmic norm

The growth of solutions of a linear system $x^\Delta = Ax$ with t-independent operator A depends on the spectrum of A. In case of a differential equation thus the maximum of the real parts of the eigenvalues of A plays a vital role. In the difference equation case the maximal modulus of the eigenvalues is crucial. Our calculus covers the full range between these two extrems and the question arises which is the appropriate description of the relation between the spectrum of A and the growth of solutions in the general context. An answer is given by the notion of logarithmic norm.

Let $\mathcal{B}$ be a Banach algebra with unity. For each $h \geq 0$ the logarithmic norm $\chi_h : \mathcal{B} \to \mathbb{R}$ is then defined to be (cf. Lakshmikantham/Leela [4], p.104)

$$\chi_h(A) := \lim_{k\to h+0} \frac{|Ak+1| - 1}{k}. \tag{7.1}$$

Without proof we state some elementary properties of the logarithmic norm.

Theorem 10: For $A, B \in \mathcal{B}$ and $h, k \in \mathbb{R}_0^+$ we get:

(i) $h \leq k \Rightarrow \chi_h(A) \leq \chi_k(A)$,

(ii) $|\chi_h(A)| \leq |A|$,

(iii) $|\chi_h(A) - \chi_h(B)| \leq |A - B|$,

(iv) $Ah + 1 \neq 0 \Rightarrow \chi_h(A) \cdot h + 1 > 0$,

(v) The mapping $\chi : \begin{cases} \mathbb{R}_0^+ \times \mathcal{B} \to \mathbb{R} \\ (h, A) \mapsto \chi_h(A) \end{cases}$ is continuous.

Remarks: The statements (i) and (ii) show that in case $h > 0$ the logarithmic norm of A is $\frac{|Ah+1|-1}{h}$. (i) and (iv) are easily seen to be true. (v) follows from (iii) and the fact that $\chi_h(A)$ ist continuous in h for fixed A.

The logarithmic norm appears later in this paper in connection with the fundamental estimate for the transition operator. That's why - in two particular cases - we give a better estimate than that in (ii).
a) For $\mathcal{B} := \mathbb{C}$ and $h > 0$ $\chi_h(\cdot)$ maps points z from $\mathbb{C}$ to the real axis in the following way: Draw a circle with center $-\frac{1}{h}$ through z, then $\chi_h(z)$ is the point on the real interval $(-\frac{1}{h}, \infty)$ where the circle meets this interval. In particular one gets $\lim_{h \to 0} \chi_h(z) = Re\, z$.
b) If X is a finite-dimensional Banach space and $\mathcal{B} := \mathcal{B}(X)$ the corresponding matrix algebra, then the following statement can be shown to follow e.g. from Stoer/Bulirsch [5].

Theorem 11: Let $A \in \mathcal{B}(X)$ be fixed. Then for any $\varepsilon \geq 0$ the exists an equivalent norm $|\cdot|_\varepsilon$ on X, such that for all $h \geq 0$ one gets for the corresponding logarithmic norm $\chi_{h,\varepsilon}$ the estimate

$$\chi_{h,\varepsilon}(A) \leq max \chi_h(\Sigma(A)) + \varepsilon \tag{7.2}$$

where $\Sigma(A)$ denotes the spectrum of A. In case every eigenvalue of A is semisimple then ε can be chosen to be 0.

8. Qualitative properties of linear dynamic processes

Again let $\mathbf{T}$ be a time scale, X be a real Banach space and $\mathcal{B}(X)$ the Banach algebra of linear continuous operators from X to X. Furthermore let $A : \mathbf{T}^\kappa \to \mathcal{B}(X)$ be a rd-continuous and regressive mapping. Then we define the mapping

$$\alpha : \begin{cases} \mathbf{T}^\kappa \to \mathbb{R} \\ t \mapsto \chi_{\mu^*(t)}(A(t)) \end{cases} \tag{8.1}$$

and consider the scalar dynamic equation

$$x^\Delta(t) = \alpha(t)x(t) \tag{8.2}$$

on $\mathbf{T}$. By means of (iv) and (v) of Theorem 10 it can be seen that α is rd-continuous and regressive. In order to emphasize the scalar character of equation (8.2) we denote the fundamental solution and the transition operator of (8.2) by

$$e_\alpha(\cdot, \tau) : \mathbf{T} \to \mathbb{R} \text{ and } e_\alpha(t, \tau) \in \mathbb{R}, \text{ respectively.}$$

The relevance of the definition (8.1) and the logarithmic norm can be expressed as follows:

Theorem 12: For any $t, \tau \in \mathbf{T}$ one has

$$|\Phi_A(t, \tau)| \leq e_\alpha(t, \tau) \text{ if } t \geq \tau. \tag{8.3}$$

Remark: If we set

$$\beta(t) := \tilde{-}\chi_{\mu^*(t)}(\tilde{-}A(t)), \tag{8.4}$$

then from (8.3) and Theorem 9, (i), (iii) we obtain the familiar looking relation

$$|\Phi_A(t,\tau)| = |\Phi_{\tilde{-}A}(\tau,t)| \le e_{\tilde{-}\beta}(\tau,t) = e_\beta(t,\tau) \text{ if } t \le \tau. \tag{8.5}$$

We want to give a complete proof of Theorem 12 in order to exhibit an example of a typical proof within the context of dynamic processes on inhomogeneous time scales (cf. Remark 1 to Theorem 5).

Proof of Theorem 12: (1) Let $\varepsilon : \mathbf{T}^\kappa \to \mathbb{R}^+$ be an arbitrary rd-continuous function. We will show by mean of the induction principle (Section 2) that for fixed $\tau \in \mathbf{T}$ the statement

$$A(t): \quad |\Phi_A(t,\tau)| \le e_\alpha(t,\tau) \cdot e_\varepsilon(t,\tau)$$

is true for all $t \in \mathbf{T}, t \ge \tau$.

Obviously the condition (I) and (IV) of Theorem 1 are satisfied.

Ad(II): $|\Phi_A(\sigma(t),\tau)| \le |A(t)\mu^*(t) + I|\ |\Phi_A(t,\tau)| = (\alpha(t)\mu^*(t)+1)|\Phi_A(t,\tau)| \le (\alpha(t)\mu^*(t)+1)(\varepsilon(t)\mu^*(t)+1)|\Phi_A(t,\tau)| \le e_\alpha(\sigma(t),t)\cdot e_\varepsilon(\sigma(t),t)\cdot e_\alpha(t,\tau)\cdot e_\varepsilon(t,\tau) = e_\alpha(\sigma(t),\tau)\cdot e_\varepsilon(\sigma(t),\tau)$.

Ad(III): Let t be right-dense. Then there exists a neighborhood U of t, such that for all $s \in U$:

$$\left|\Phi_A(s,t) - \Phi_A(t,t) - A(t)\Phi_A(t,t)(s-t)\right| \le \frac{\varepsilon(t)}{3}|s-t|,$$

$$\left||A(t)(s-t)+1| - (1+\alpha(t)(s-t)\right| \le \frac{\varepsilon(t)}{3}|s-t|,$$

$$|e_{\alpha\tilde{+}\varepsilon}(s,t) - e_{\alpha\tilde{+}\varepsilon}(t,t) - (\alpha\tilde{+}\varepsilon)(t)e_{\alpha\tilde{+}\varepsilon}(t,t)(s-t)| \le \frac{\varepsilon(t)}{3}|s-t|. \tag{8.6}$$

Now, for $s \in U, s \ge t$, we have:

$$\begin{aligned}
&|\Phi_A(s,t)| \le |\Phi_A(s,t) - \Phi_A(t,t) - A(t)\Phi_A(t,t)(s-t)| + \\
&+|\ |1+A(t)(s-t)| - (1+\alpha(t)(s-t))| + (1+\alpha(t)(s-t)) \le \\
&1 + [\alpha(t) + \alpha(t)\varepsilon(t)\mu^*(t) + \tfrac{2}{3}\varepsilon(t)](s-t) = \\
&1 + (\alpha\tilde{+}\varepsilon)(t)\cdot(s-t) - \tfrac{\varepsilon(t)}{3}(s-t) \le e_{\alpha\tilde{+}\varepsilon}(s,t).
\end{aligned}$$

The last inequality follows from (8.6).

From the validity of $A(t)$ we get the validity of $A(s)$:

$$|\Phi_A(s,\tau)| \le |\Phi_A(s,t)||\Phi_A(t,\tau)| \le e_{\alpha\tilde{+}\varepsilon}(s,t)e_{\alpha\tilde{+}\varepsilon}(t,\tau) = e_{\alpha\tilde{+}\varepsilon}(s,\tau).$$

(2) It remains to be proved that $\lim_{n\to\infty} e_{1/n}(t,\tau) = 1$ for fixed $t \in \mathbf{T}$.
For $n \in \mathbb{N}$ we have

$$\chi_{\mu^*(t)}(1/n) = 1/n \text{ and } (2\tilde{-}1/n)(t) = \frac{2-1/n}{\mu^*(t)/n+1} > 0.$$

From the validity of $A(t)$ we get for all $t \geq \tau$:

$$|e_{1/n}(t,\tau)| \leq e_{1/n}(t,\tau) \cdot e_{2\tilde{-}1/n}(t,\tau) = e_2(t,\tau).$$

With the mean value theorem (Theorem 5) it follows that

$$|e_{1/n}(t,\tau) - 1| \leq \sup_{s\in[\tau,t]} |1/n \cdot e_{1/n}(s,\tau)||t-\tau| \leq 1/n \cdot e_2(t,\tau)(t-\tau).$$

For fixed $t \geq \tau$ we obtain: $\lim_{n\to\infty} e_{1/n}(t,\tau) = 1$.
This completes the proof of Theorem 12.

By means of the estimates (8.3) and (8.5) we may draw conclusions about the qualitative behavior of solutions of linear dynamic equations. W.l.o.g. we consider the direction of increasing time only. If we suppose that the time scale $\mathbf{T}$ is unbounded above in $\mathbb{R}$ then we can show the following: If $\beta : \mathbf{T} \to \mathbb{R}$ is positive and bounded away from zero then one gets

$$\lim_{t\to\infty} e_\beta(t,\tau) = \infty.$$

If $\alpha : \mathbf{T} \to \mathbb{R}$ is negative and bounded away from zero we get

$$\lim_{t\to\infty} e_\alpha(t,\tau) = 0.$$

At this point it is possible to introduce the notion of stability and to start stability theory for linear dynamic processes. We cannot do so in this paper, we just mention one basic result which follows from Theorem 12. To this end we define for each $h \geq 0$ the generalized closed disk with center $-\frac{1}{h}$ and radius $\frac{1}{h}$

$$S_h := \begin{cases} \{z \in \mathbb{C} : |z + \frac{1}{h}| \leq \frac{1}{h}\} & \text{if } h > 0 \\ \{z \in \mathbb{C} : Re\, z \leq 0\} & \text{if } h = 0 \end{cases}.$$

Theorem 13: Let X be finite-dimensional and suppose
(i) $h := sup\{\mu^*(t) : t \in \mathbf{T}^\kappa\} < \infty$,
(ii) $A \in \mathcal{B}(X)$ is regressive,
(iii) all eigenvalues of A are in S_h,
(iv) on the boundary of S_h there are only semisimple eigenvalues.
Then any solution of $x^\Delta = Ax$ is bounded as t increases.

This theorem apparently contains well known results for the special cases of differential equations $\frac{dx}{dt}(t) = Ax(t)$ and difference equations $x(t+1) - x(t) = Ax(t)$, i.e.

$x(t+1) = (A+I)x(t)$. We finally wish to emphasize that our calculus provides a smooth transition from the discrete case where the "stability region" is the shifted unit disk to the continuous case with the left half plane as "stability region".

References

[1] Aulbach, B. Hilger S.: *A Unified Approach to Continuous and Discrete Dynamics, in "Differential Equations: Qualitative Theory"*. Colloq. Math. Soc. János Bolya:, 48. North Holland, to appear.

[2] Dieudonné, J.: *Foundations of Modern Analysis* Academic Press, New York-London, 1969

[3] Hilger, S.: *Der Maßkettenkalkül - ein vereinheitlichter Zugang zu Differential- und Differenzenrechnung*. Preprint Nr. 213, Univ. Augsburg, 1989

[4] Lakshmikantham, V., Leela, S.: *Differential and Integral Inequalities* Academic Press, New York - London, 1969

[5] Stoer, J., Bulirsch, R.:Introduction to Numerical Analysis *Springer, New York u.a., 1983*

BIFURCATIONS AND ATTRACTORS OF PHASE SYSTEMS

V.N. Belykh [1)]

Consider the dynamical systems

$$\ddot{x}_i + X_i(x_i,\dot{x}_i) = f_i(x,\dot{x},y)$$
$$\dot{y} = Y(x,\dot{x},y), \qquad i = \overline{1,n}, \tag{1}$$

where $U = \|u_i\|_n^1$, $U = (x,X,f)$ are n-vectors and $V = \|v_j\|_m^1$, $V = (y,Y)$ are m-vectors.

The systems describe the dynamics of various physical systems, namely, the dynamics of interacting oscillators and rotators, of interconnected phase synchronization systems, of Josephson junctions and pendulums, of controlled electromechanical systems and other systems. Systems (1) are called phase systems when $x \in T^n(x_i \in S)$.

In the process of investigation of particular systems the bifurcation problem appears. This is a problem about partitioning of parameters space into domains which correspond to various qualitative portraits of system (1) in phase space. The phase portraits differ from each other in attractors, i.e. limit sets of trajectories (stationary, periodical, homoclinic, etc.) and in mutual disposition of their stable and unstable manifolds (separatrices, domains of attraction and of repulsion). For short we shall not distinguish the notions of attractor and repeller (i.e. the attractor as $t \to -\infty$).

Here the two-dimensional comparison systems method [1 - 4] proved to be fruitful. This method provides the opportunity to determine not only the attractors and their attraction domains but also nonlocal bifurcations of separatrix manifolds. So it provides the opportunity to establish the existence of complicated structures of trajectories.

As soon as further we shall discuss the results of developement of this method will briefly review here its essence. We consider a particular case when (1) is dissipative with respect to y, which means that the trajectories of (1) do not leave the domain $g_\alpha = \{\|y\| < \ell\}$, are bounded in the space $(x,\dot{x})$ and $g_\alpha : K_i^- < f_i < K_i^+$. Let us introduce 2n equations of second order

$$\ddot{x}_i + X_i(x_i,\dot{x}_i) = K_i^+, \quad \ddot{x}_i + X_i(x_i,\dot{x}_i) = K_i^- \tag{2}$$

and denote by A_i^+ and A_i^- the two-dimensional systems (with $\dot{x}_i = v_i$) which correspond to (2). Let us also introduce 2n glued systems $\tilde{A}_i^{+(-)} = \{A_i^{+(-)}$ when $v_i \geq 0$, $A_i^{-(+)}$ when $v_i < 0\}$. If the function

1) Institute of Water Transport Engineers, Head of Department of Mathematics, 603 000 Gorky, Gorky Str. 144, USSR

$X_i(x_i,0)$ is monotonically increasing and has a simple zero on the interval (a_i,b_i) and the dissipation conditions $(X_i)'_{v_i} > \lambda > 0$ are fulfilled (λ satisfies certain conditions [1,2]), then the system $\tilde{A}_i^+$ has a stable limit cycle, which bounds the <u>attracting domain</u> g_i^+, i.e. the domain that absorbs the trajectories from (a_i,b_i).
Similary in case $(X_i)'_{v_i} < -\lambda < 0$ there exists a repelling domain g_i^- that is bounded by unstable cycle of $\tilde{A}_i^-$. If function $X_i(x_i,0)$ is monotonically decreasing and has a simple zero on (a_i,b_i) then the saddle separatrices of systems A_i^+ and A_i^- (independently of $(X_i)'_{v_i}$) form the <u>saddle cell</u> $g_i^o = g_i^s \cap g_i^u$ where g_i^s and g_i^u are the <u>separatrix beds</u>. Introduce the notation $* = ((0) \vee (+) \vee (-))$.

<u>Assertion:</u> If the domains g_i^* do not leave the strips on (x_i,v_i) which correspond to the intervals (a_i,b_i), then the projections of the attractors (for the trajectories from these strips) on to the plane (x_i,v_i) are in g_i^* and the projections of separatrices manifolds are in $g^{s(u)}$.

Generally speaking, the question about the structure of attractors is left open. It may be treated in different ways. Thus, for a second order equation

$$\ddot{x} + X(x,\dot{x},t) = 0$$

the conditions of uniqueness of attracting, repelling and saddle trajectories (either periodical or bounded at both sides) in the domains g^* coincide with the conditions of existence of these domains. In other particular cases (one nonlinearity Hurwitz property, etc.) it is possible to get these conditions with the help of Ljapunov functions, by means of the Leonov reduction method [5], by means of frequency methods, etc. We must notice that in the case when the attractor of system (1) is a unique bounded trajectory in the corresponding domain of phase space, that is in the topological product $G = \prod_{i=1}^{n} g_i^*$, is an essential one. Really, if in this connection all the $*_i = (+)$ then the global stability takes place in G and if among g_i^* there is one saddle cell $g_{i_1}^o$ at least then there exists in G a saddle trajectory, the separatrices of which are disposed in the beds $g_{i_1}^{s(u)}$. Thus in the latter case there appears the possibility to establish the existence of bifurcation of separatrices and consequently, the existence of homoclinic trajectories and consequently, the existence of complicated structures.

We give further the topological conditions of global stability and dichotomy of system (1). In this connection we define the domains g_i^s and g_i^u that majorize the separatrix manifolds.

Suppose system (1) satisfies the following conditions.

1) It has a zero equilibrium state O, i.e.

$$X(0,0) = f(0,0,0) = Y(0,0,0) = 0.$$

2) The system $\dot{y} = \bar{Y}(x,\dot{x},y)$, where

$$\bar{Y} = \begin{cases} Y(x,\dot{x},y), & x,\dot{x} \in D = \{|x_i| < \alpha_i,\ |\dot{x}_i| < \beta_i,\ i = \overline{1,n}\} \\ Y(\alpha,\beta,y), & x,\dot{x} \notin D, \end{cases}$$

$\alpha = \|\alpha_i\|_n^1$, $\beta = \|\beta_i\|_n^1$ are constant vectors, is dissipative in the domain $G_\gamma = \{\|y\| < \gamma(\alpha,\beta)\}$ such that $\gamma \xrightarrow[\alpha\to 0,\beta\to 0]{} 0$.

3) The two-dimensional comparison systems

$$\ddot{x}_i + X_i(x_i,\dot{x}_i) = f_i^{+(-)}, \quad \dot{x}_i = v_i, \quad i = \overline{1,n} \quad (A_i^{+(-)})$$

where $f_i(\alpha,\beta,\gamma) = f_i^+ > f_i^- = f_i(-\alpha,-\beta,-\gamma)$ have the following properties: For all i either the system $\tilde{A}_i^{+(-)}$ has a domain $g_i^{+(-)}$ or $A_i^{+(-)}$ forms a saddle cell g_i^o.

<u>Theorem.</u> Suppose conditions 1) - 3) hold. Suppose also there exists at least one path along which vectors α and β tend to zero simultaneously and

$$g_i^* \subset D = \{|x_i| < \alpha_i,\ |\dot{x}_i| < \beta_i,\ i = \overline{1,n}\}. \qquad (!)$$

If all $\tilde{A}_i^+$ have domains g_i^+ ($*_i = (+), \forall i$) then the system (1) is absolutely stable. If among g_i^* there are $g_{i_1}^+$ and $g_{i_2}^-$ or among g_i^* there is at least one saddle cell ($\exists *_i = (0)$) then the system (1) is dichotomous. In the latter case the O-saddle and its separatrices are disposed in the cones which adjoin the point O and are formed by the edges of the segment $g_i^{s(u)} \cap \{x_i = \alpha_i\}$.

The proof of the Theorem is based on the fact that under the hypotheses of the Theorem the trajectories of system

$$\ddot{x}_i + X_i(x_i,\dot{x}_i) = \bar{f}_i(x,\dot{x},\gamma)$$
$$\dot{y} = \bar{Y}(x,\dot{x},y),$$

where

$$\bar{f}_i = \begin{cases} f_i(x,\dot{x},y), & x,\dot{x} \in D, \\ f_i(\alpha,\beta,\gamma), & x,\dot{x} \notin D, \end{cases}$$

coincide with the trajectories of system (1) in the domain $D \times G_\gamma$.

This Theorem without any changes is valid for a nonautonomous system (1), i.e. for the case when f and g depend upon time. If this dependence is a periodical one then O plays the part of periodical motion.

In case system (1) has several equilibrium states. The investigation should be done separately for every domain where function $X_i(x_i,0)$ is monotonous. In this case conic domains which majorize the separatrices must be continued outside the domains of monotonicity by the separatrix beds $g_i^{s(u)}$ of system (2).

Example 1. Consider the system

$$\begin{aligned} \ddot{\varphi}_1 + \lambda_1\dot{\varphi}_1 + a_{11} \sin \varphi_1 + a_{12} \sin \varphi_2 &= \gamma_1, \\ \ddot{\varphi}_2 + \lambda_2\dot{\varphi}_2 + a_{21} \sin \varphi_1 + a_{22} \sin \varphi_2 &= \gamma_2. \end{aligned} \tag{4}$$

Its equilibrium states are defined by equations

$$\sin \varphi_1 = \gamma_1^* = \frac{1}{\Delta} (\gamma_1 a_{22} - \gamma_2 a_{12});$$

$$\sin \varphi_2 = \gamma_2^* = \frac{1}{\Delta} (\gamma_2 a_{11} - \gamma_1 a_{21}).$$

When $|\gamma_{1,2}^*| < 1$ there are four equilibrium states which correspond to four domains of monotonicity of functions $\sin \varphi_1$ and $\sin \varphi_2$. Consider the domain $S_1 = \{-\frac{\pi}{2} < \varphi_{1,2} < \frac{\pi}{2}\}$. Substituting $\sin \varphi_2$ instead of $\gamma_2^* \pm \delta$ in the first equation and $\sin \varphi_1$ instead of $\gamma_1^* \pm \delta$ in the second equation we get the systems

$$\ddot{\varphi}_1 + \lambda_1\dot{\varphi}_1 + a_{11} \sin \varphi_1 = a_{11}\gamma_1^* \pm a_{12}\delta \qquad (A_1^{+(-)})$$

$$\ddot{\varphi}_2 + \lambda_2\dot{\varphi}_2 + a_{22} \sin \varphi_2 = a_{22}\gamma_2^* \pm a_{21}\delta \qquad (A_2^{+(-)}).$$

In this connection $\alpha_{1,2} = \arcsin(\gamma_{1,2}^*) - \arcsin(\gamma_{1,2}^* - \delta)$. Suppose for simplicity that $\lambda_{1,2} > \lambda_{1,2}^*$, where $\lambda_{1,2}^*$ correspond to stable knots (not foci) of systems $A_{1,2}^{+(-)}$ ($\lambda_{1,2}^* \leq 2$). Then the attracting domains $g_{1,2}^+$ are located on cylinders $(\varphi_{1,2}, \dot{\varphi}_{1,2})$, in the band between these stable knots and consequently the conditions (!) take the following form

$$\frac{a_{12}}{a_{11}} < 1, \qquad \frac{a_{21}}{a_{22}} < 1, \tag{5}$$

which means that the system (4), (5) is asymptotically stable in the domain S_1. It is not difficult to check that (5) is the condition for three other equilibrium states to be the saddles and for the system (4) to be dichotomous in the corresponding domains. Let us consinder further the mutual disposition of separatrix beds $g_{1,2}^{s(u)}$ for the systems $A_{1,2}^{+(-)}$. We receive that under the additional condition

$$|\gamma_1^*| < \gamma_T \left(\frac{\lambda_1}{\sqrt{a_{11}}}\right) - \frac{a_{12}}{a_{11}} \delta,$$

$$|\gamma_2^*| < \gamma_T \left(\frac{\lambda_2}{\sqrt{a_{22}}}\right) - \frac{a_{21}}{a_{22}} \delta,$$

where $\gamma = \gamma_T(\lambda)$ is the Tricomi curve the system (4) is globally asymptotically stable in the whole phase space. With the help of mutual disposition of separatrix beds we may state that system (4) has Shilnikov spiral attractors, linked with befurcations of - the loops of saddle - foci separatrices which correspond to the domains of different monotonicity of functions sin $\varphi_{1,2}$.

Example 2. Consider the system

$$c\ddot{\varphi} - \dot{\varphi} - \sin\varphi = \gamma + x + a\dot{x},$$
$$\ddot{x} + \delta\dot{x} + \omega^2 x = \delta\ddot{\varphi} + d\dot{\varphi}, \tag{6}$$

which deseribes the following system: Josephson junction - RLC - circuit [6]. With the help of two-dimensional comparison systems method we can determine bifurcations and various attractors (including the spiral one) of system (6) in the domain $|\gamma| < 1$, when (6) has an equilibrium state. In domain $|\gamma| > 1$ the system (6) has no equilibrium states and the comparison method establishes only the existence of rotating motions. In this case for $c^{-1} = \mu \ll 1$ with the help of substitution

$$\left\| \begin{matrix} x \\ \frac{\dot{x}}{\omega} \end{matrix} \right\| = R(\varphi) \cdot \left\| \begin{matrix} \theta \\ \eta \end{matrix} \right\| + \left\| \begin{matrix} \frac{d}{\omega} \\ -b \end{matrix} \right\|$$
$$\dot{\varphi} = \omega + \mu(\Phi(\theta,\eta,\varphi) + y)$$

(where $R(\varphi)$ is a matrix of rotation about the angle φ, Φ is a periodical with respect to φ function), averaging and the subsequent linear substitution we can reduce the system to the following one

$$\dot{x} = -z(x-y) + \varrho$$
$$\dot{y} = -y + rx - zx \tag{7}$$
$$\dot{z} = -z + xy + \lambda x.$$

As soon as the system (7) in case $\varrho = \lambda = 0$ is the Lorenz system we thus establish the existence of a strange attractor of the Lorenz type for the rotating movements of system (6).

In conclusion we give an example of piecewise linear system from phase synchronization theory and establish that the difference sheme of Euler broken lines method for this system has the hyperbolic strange attractor close to the Belykh-attractor [3].

References

[1] Belykh, V.N.: On the qualitative investigation of a nonautonomous equation of the second order. Differentzial'nye Uravnenija, 11 (1975) 10, 1738 - 1753.

[2] Belykh, V.N.: Qualitative methods of nonlinear oscillations theory for lumped systems. Text-book. State University, Gorky 1980.

[3] Belykh, V.N.: On the qualitative structures and bifurcations of particular dynamical systems. In the book "IX International conference on nonlinear oscillations", v. 2, Qualitative methods. Kiev "Naukova Dumka", 1984, 45 - 48.

[4] Belykh, V.N.: A two-dimensional comparison systems method in the qualitative theory of particular dynamical systems. Doctorskaja dissertatzija. Gorky, State University, 1985.

[5] Gelig, A.Kh.; G.A. Leonov; V.A. Jakubovich: Stability of nonlinear systems with multiple equilibria. Nauka, Moscow 1978.

[6] Verichev, N.N.: The Lorenz equation in the system rotator-oscillator. In the book: Methods of differential equations qualitative theory. Gorky, State University, 1984, 54 - 59.

ALGEBRAIC TOPOLOGICAL ENTROPY

Thomas Hudetz[1]

Abstract

As a "by-product" of the Connes-Narnhofer-Thirring theory of dynamical entropy for (originally *non*-Abelian) nuclear C^*-algebras, the well-known variational principle for topological entropy is equivalently reformulated in purely *algebraically defined* terms for (separable) *Abelian* C^*-algebras. This "algebraic variational principle" should not only nicely illustrate the "feed-back" of methods developed for quantum dynamical systems to the classical theory, but it could also be *proved directly* by "algebraic" methods and could thus further simplify the original proof of the variational principle (at least "in principle").

1 Introduction

The notion of *topological entropy* had been introduced in 1965 by Adler, Konheim and McAndrew [1] as a conjugacy invariant of continuous transformations on compact topological spaces, by close analogy with the Kolmogorov-Sinai [14] entropy invariant of measure preserving transformations on probability spaces, and thus leading [1] to the conjecture (relating topological and measure-theoretic entropy, see section 2) which was called *variational principle* for "physical" reasons later on (cf. [21], e.g.). The first part of this conjecture (that topological entropy *bounds* measure-theoretic entropy) was first proved by Goodwyn [10] in 1968, and this inequality (2.6,i) was then sharpened to the variational *equality* (2.6,ii) by Dinaburg [6] and Goodman [9] in 1970.

Although these original proofs have been simplified *technically* due to the work of Bowen (and others), resulting in the proofs to be found in the textbooks on ergodic theory (see [5,23] and the references therein), there still remained the *conceptual* difficulty that comparing topological with measure-theoretic entropy involved a comparison of open covers (respectively, of spanning or separated sets in Bowen's equivalent definition of topological entropy) with "disjoint" Borel partitions of the compact metric space considered. This difficulty was partly overcome by Palm [20], who defined the entropy for *dynamical lattices* as a common generalization of topological and measure-theoretic entropy, such that also the latter is equivalently defined in terms of *not* necessarily *disjoint* Borel covers. This lattice-theoretic reformulation allowed a simple proof at least of Goodwyn's theorem (2.6,i); but still the proof involved a comparison of *all* Borel covers (as used in the redefinition of measure-theoretic entropy) with their "generating" subset of *open* covers, by means of a generalized Kolmogorov-Sinai theorem.

[1]Institut für Theoretische Physik, Universität Wien, Boltzmanngasse 5, A-1090 Wien

In this contribution, the (Borel-) measure-theoretic entropy of a homeomorphism $T : X \to X$ on a compact metric space X will be shown (4.1) to coincide with the *Connes-Narnhofer-Thirring* (CNT) *entropy* ([4], see section 3) of the *-automorphism induced by T on the (separable) Abelian C^*-algebra $C(X)$ of (complex-valued) continuous functions on X, what can be used as equivalent *definition* (3.6) because of the Gelfand *-isomorphism from *any* (separable) Abelian C^*-algebra $\mathcal{A} \ni \mathbf{1}$ to some $C(X)$ as above. This CNT entropy of a *-automorphism acting on $\mathcal{A} \cong C(X)$ is defined (4.2) in terms of (unity-preserving) positive linear maps from finite-dimensional Abelian C^*-algebras into $\mathcal{A}$, corresponding to finite partitions of unity $\mathbf{1} \in C(X)$ into positive continuous functions on X; and each positive continuous partition of $\mathbf{1} \in C(X)$ in turn (uniquely) determines an open cover of X. This correspondence will be used (4.3) for equivalently (4.5) redefining also the *topological* entropy of $T : X \to X$ in terms of these positive linear maps from finite-dimensional Abelian algebras into $\mathcal{A} \cong C(X)$ (with the *-automorphism induced by T), such that the variational principle (2.6) can be reformulated in purely *algebraically defined* terms for (separable) Abelian C^*-algebras $\mathcal{A}$. Furthermore, this "algebraic variational principle" (4.6) does *not* involve any more a comparison of *different* objects as arguments of the two different entropy functionals, compared for a fixed *-automorphism of $\mathcal{A}$. In spite of this *conceptual* simplification, however, the *direct* proof of the algebraic variational principle (4.6) has been hindered up to now[2] by the *technical* difficulties encountered in explicitly *calculating* the CNT-entropy for a given finite partition of unity $\mathbf{1} \in \mathcal{A}$ into positive elements from $\mathcal{A} \cong C(X)$. A first step in this direction will be made as an "appendix" to the present contribution, but the further steps have to be left to future investigations.

2 The Topological Variational Principle

Throughout this contribution (as in [5]), by a *topological dynamical system* (t.d.s.) (X, T) we shall understand a compact *metric* space X together with a *homeomorphism* $T : X \to X$. By definition, any open cover of X possesses a *finite* subcover; and we can restrict ourselves to the latter from the very beginning (cf. [5], p. 83), denoting by $\mathcal{O}(X)$ or simply $\mathcal{O}$ the set of finite open covers of X. Using the notation of [23], we recall the original definition [1] of topological entropy:

Definition (2.1): Let (X, T) be a t.d.s. We define the *join* of $\alpha, \beta \in \mathcal{O}(X)$ by $\alpha \vee \beta = \{A \cap B | A \in \alpha, B \in \beta\}$ and the *action* of T on $\alpha \in \mathcal{O}$ by $T^{-1}\alpha = \{T^{-1}A | A \in \alpha\}$, and we use the short notation $\alpha_T^n = \bigvee_{i=0}^{n-1} T^{-i}\alpha$ $(n \in N)$.

[2] Deadline for the proceedings: November 15, 1989.

(i) We denote by $N(\alpha)$ the number of sets in a *minimal* subcover of $\alpha \in \mathcal{O}$ (with minimal cardinality). The *entropy* of $\alpha \in \mathcal{O}$ is defined by $H(\alpha) = \log N(\alpha)$.

(ii) The "entropy" of T w.r.t. $\alpha \in \mathcal{O}$ is defined by $h(T,\alpha) = \lim_{n\to\infty} \frac{1}{n} H(\alpha_T^n)$.

(iii) The *topological entropy* of T is defined by $h(T) = \sup_{\alpha\in\mathcal{O}} h(T,\alpha)$.

The basic properties of topological entropy are its *conjugacy invariance* $h(T) = h(\Phi \circ T \circ \Phi^{-1})$ for a t.d.s. (X,T) with a homeomorphism $\Phi : X \to X'$ (X' compact metric), and its *scaling* property $h(T^k) = |k| \cdot h(T) \ \forall\ k \in \mathbf{Z}$. – The analogous dependence on T is shown also by the Kolmogorov-Sinai [14] *measure-theoretic* entropy $h_\mu(T)$ of T as Borel measurable transformation, but in addition $h_\mu(T)$ depends on the T-invariant Borel probability measure $\mu = \mu \circ T^{-1}$ on X, chosen from the non-empty convex set $M(X,T)$ of all such measures. The original measure-theoretic definition (cf. [23]) of $h_\mu(T)$ is given in terms of the set $\mathcal{P}(X,\mu)$ of finite Borel partitions (μ-mod 0) of X (or the corresponding finite σ-algebras of Borel sets); and because of the obvious one-to-one correspondence (cf. the proof of (2.3) below) between $\mathcal{P}(X,\mu)$ and the set $\mathcal{F}(X,\mu)$ of all finite-dimensional *-subalgebras (with unity $\mathbf{1}$) of the Abelian W^*-algebra $\mathcal{M}_\mu \equiv L^\infty(X,\mu)$, we can equivalently reformulate the definition in *algebraic* terms (cf. [13,12]):

Definition (2.2): Let (X,T) be a t.d.s. and $\mu \in M(X,T)$. We define a **-automorphism* τ_μ resp. a (faithful, normal) *state* ω_μ on $\mathcal{M}_\mu \equiv L^\infty(X,\mu)$ by $\tau_\mu(f) = f \circ T$, resp. $\omega_\mu(f) = \int_X f(x)d\mu(x) \ \forall\ f \in \mathcal{M}_\mu$; such that $\omega_\mu \circ \tau_\mu = \omega_\mu$. We denote by $\mathcal{F}$ the set of all finite-dimensional unital *-subalgebras of $\mathcal{M}_\mu$, and $\mathcal{A} \vee \mathcal{B}$ ($\in \mathcal{F}$) denotes the subalgebra *generated* by $\mathcal{A}, \mathcal{B} \in \mathcal{F}$.

(i) For $\mathcal{A} \in \mathcal{F}$, $\dim \mathcal{A} = n \in \mathbf{N}$, with (uniquely determined, cf. [22], p. 50/51) minimal projectors $P_i \in \mathcal{A}$ ($i = 1,\ldots,n$), the *entropy* of $\mathcal{A}$ is defined by

$$H_{\omega_\mu}(\mathcal{A}) = \sum_{i=1}^{n} \eta(\omega_\mu(P_i)),$$

where $\eta(x) \equiv -x\log x \ \forall\ x \in [0,1]$ with $\eta(0) \equiv 0$.

(ii) The "entropy" of τ_μ w.r.t. $\mathcal{A} \in \mathcal{F}$ is defined by

$$h_{\omega_\mu}(\tau_\mu, \mathcal{A}) = \lim_{n\to\infty} \frac{1}{n} H_{\omega_\mu}(\mathcal{A} \vee \tau_\mu\mathcal{A} \vee \ldots \vee \tau_\mu^{n-1}\mathcal{A}).$$

(iii) The "Kolmogorov-Sinai" *entropy* of τ_μ is defined by

$$h_{\omega_\mu}(\tau_\mu) = \sup_{\mathcal{A}\in\mathcal{F}} h_{\omega_\mu}(\tau_\mu, \mathcal{A}).$$

Proposition (2.3): Let (X, T) be a t.d.s. and $\mu \in M(X, T)$, then the entropy (2.2,iii) coincides with the (original) *measure-theoretic* entropy $h_\mu(T)$ of T w.r.t. μ (cf. [23,5]): $h_{\omega_\mu}(\tau_\mu) = h_\mu(T)$.

Proof: Obvious from Def. (2.2) because of the one-to-one correspondence between $\xi \in \mathcal{P}(X, \mu)$ and $\mathcal{A}(\xi) \in \mathcal{F}(X, \mu)$: To $\xi = \{A_1, \ldots, A_n\}$ there corresponds a finite partition of unity $\mathbf{1} = \sum_{i=1}^n P_i$ in $\mathcal{M}_\mu \equiv L^\infty(X, \mu)$ given by the (μ-equivalence classes of) characteristic functions $\chi_{A_i} \equiv P_i$; and we denote by $\mathcal{A}(\xi) \subset \mathcal{M}_\mu$ the n-dimensional *-subalgebra generated by the P_i $(i = 1, \ldots, n)$, which in turn give the uniquely determined minimal projectors $P_i \in \mathcal{A}(\xi)$.

Definition (2.4): The *entropy map* of a t.d.s. (X, T) is the map $h_T : M(X, T) \to [0, \infty]$ defined by $h_T(\mu) = h_{\omega_\mu}(\tau_\mu)$ $\forall$ $\mu \in M(X, T)$, with Def. (2.2,iii) for the r.h.s.

Remarks (2.5):

1. Together with Prop. (2.3) above, this definition of the entropy map coincides with that of [23], simply expressing the dependence of $h_\mu(T)$ on $\mu \in M(X, T)$. "Strictly" speaking, however, the value of the entropy map (2.4) at $\mu \in M(X, T)$ has to be computed for the Abelian W^*-algebra $\mathcal{M}_\mu \equiv L^\infty(X, \mu)$ with *-automorphism τ_μ (both dependent on μ); and this "abstract nonsense" shows that the W^*-algebraic reformulation (2.2) is not well suited for defining an entropy map, the reason being that in general $M(X, T)$ is not contained in any dual $\mathcal{M}_\mu^*$ (cf. [22], p. 117) for $\mu \in M(X, T)$.

2. For the (Abelian) C^*-algebra $\mathcal{A} = C(X)$, however, $M(X, T)$ is identified with a weak-*-compact, convex (non-empty) subset of the unit ball in $\mathcal{A}^*$ (cf. [5,23], e.g.) by the bijection between the measure $\mu \in M(X, T)$ and the integral state $\omega_\mu \in \mathcal{A}^*$ as in (2.2); and the *-automorphism τ on $\mathcal{A}$, induced by T as in (2.2), is *independent* of μ. $\mathcal{A}$ would thus be taylor-made for defining an "algebraic" entropy map; but for a compact metric space X there will in general exist no finite-dimensional *-subalgebras of $\mathcal{A} = C(X)$ at all, such that the set $\mathcal{F}$ in Def. (2.2) and thus the definition itself will be empty.

3. Nevertheless, because of Prop. (2.3) also the entropy map h_T from (2.4) is *affine* on $M(X, T)$; and $h_T : M(X, T) \to [0, \infty)$ is weak-*-upper-semicontinuous for an ("asymptotically entropy-") expansive homeomorphism T (cf. [5,23]). We can also use h_T for finally recalling the *variational principle* for topological entropy:

Theorem (2.6): Let (X, T) be a t.d.s., then we have with Def. (2.1) resp. (2.4):

(i) $h(T) \geq h_T(\mu) \; \forall \; \mu \in M(X,T)$ (Goodwyn [10]), and even

(ii) $h(T) = \sup_{\mu \in M(X,T)} h_T(\mu)$ (Goodman [9]).

3 The Connes-Narnhofer-Thirring Entropy

This third section could also be read as a continuation of section 2 in the recent review [13], to which we refer for a motivation (cf. also [12]) of the following definition given by Connes, Narnhofer and Thirring [4]:

Definition (3.1): Let $\mathcal{A} \ni \mathbf{1}$ be a C^*-algebra with unity $\mathbf{1}$ and with a state ω on $\mathcal{A}$. We denote by $CP_1(\mathcal{A})$ or simply CP_1 the set of all *completely positive, unital* (c.p.u.) linear maps from finite-dimensional unital C^*-algebras into $\mathcal{A}$.

(i) For $\gamma_1, \ldots, \gamma_n \in CP_1(\mathcal{A})$ with $\gamma_k : \mathcal{A}_k \to \mathcal{A}$ $(k = 1, \ldots, n)$, an *Abelian model* is given by a c.p.u. continuous linear map $P : \mathcal{A} \to \mathcal{B} \equiv \bigvee_{k=1}^{n} \mathcal{B}_k$ from $\mathcal{A}$ onto a finite-dimensional Abelian C^*-algebra $\mathcal{B} \ni \mathbf{1}$ with subalgebras $\mathcal{B}_k \ni \mathbf{1}$ generating $\mathcal{B}$, and with a state μ on $\mathcal{B}$ such that the dual P^* of P maps μ to $\omega = P^*(\mu) \equiv \mu \circ P$.

(ii) For an Abelian model as in (i), we denote by $Q^k_{i_k}$ $(i_k = 1, \ldots, \dim \mathcal{B}_k)$ the minimal projectors in $\mathcal{B}_k$ and by $E_k : \mathcal{B} \to \mathcal{B}_k$ the c.p.u. canonical *conditional expectation* onto $\mathcal{B}_k$ with restricted state $\mu_k \equiv \mu_{|\mathcal{B}_k}$ (such that $\mu_k \circ E_k = \mu$), and we define the c.p.u. "model map" $\rho_k : \mathcal{A}_k \to \mathcal{B}_k$ by $\rho_k = E_k \circ P \circ \gamma_k$ $(k = 1, \ldots, n)$.

(iii) The *entropy* of $\gamma_1, \ldots, \gamma_n \in CP_1(\mathcal{A})$ is defined by

$$H_\omega(\gamma_1, \ldots, \gamma_n) = \sup_{\{\mu \circ P = \omega\}} [S(\mu) - \sum_{k=1}^{n} S(\mu_k) + \sum_{k=1}^{n} \sum_{i_k} \mu_k(Q^k_{i_k}) S(\mu_k \circ \rho_k | Q^k_{i_k} \circ \rho_k)],$$

where the supremum is taken over all Abelian models for $\gamma_1, \ldots, \gamma_n$. Here $S(\cdot)$ resp. $S(\cdot|\cdot)$ denotes the *entropy* resp. *relative entropy* functional of states on (Abelian, resp. non-Abelian) *finite*-dimensional (C^*-) algebras (cf. [16], also [12,13]); and in the second arguments of the relative entropy terms we have identified $\mathcal{B}^*_k$ with $\mathcal{B}_k$ defining the states $Q^k_{i_k}$ on $\mathcal{B}_k$ by $Q^k_{i_k}(Q^k_{j_k}) = \delta_{i_k j_k}$ $(i_k, j_k = 1, \ldots, \dim \mathcal{B}_k)$.

Remarks (3.2):

1. For the definition of completely positive maps resp. conditional expectations we refer e.g. to [22] (cf. also [4,12]). As remarked in [4], for any finite-dimensional C^*-algebra $\mathcal{A}_k$ there exists a matrix algebra $M_{d_k}(\mathbf{C})$ containing $\mathcal{A}_k$ as a subalgebra, together with a conditional expectation (i.e. norm-one projection) $\theta_k : M_{d_k} \rightarrow \mathcal{A}_k$; and because of the *invariance* of the *relative* entropy w.r.t. extending *both* states by θ_k (cf. [4], also [19]) we have $H_\omega(\gamma_1, \ldots, \gamma_n) = H_\omega(\gamma_1 \circ \theta_1, \ldots, \gamma_n \circ \theta_n)$ with c.p.u. maps $\gamma_k \circ \theta_k : M_{d_k} \rightarrow \mathcal{A}$. Thus we may assume $\mathcal{A}_k = M_{d_k}$. By the results of Choi and Effros (cf. [2,7]) there is a natural bijection between the set of c.p.u. maps $\gamma : M_d(\mathbf{C}) \rightarrow \mathcal{A}$ and the set $\{A = (a_{ij}) \in M_d(\mathcal{A})^+ | \sum_{i=1}^d a_{ii} = 1 \in \mathcal{A}\} \neq \emptyset$; while in general $\mathcal{A}$ does not contain any *finite*-dimensional *sub*algebras, see (2.5,2).

2. An Abelian model (i) for $n = 1$, given by a c.p.u. *continuous* linear map $P : \mathcal{A} \rightarrow \mathcal{B}$ with a state μ on $\mathcal{B}$ such that $\mu \circ P = \omega$ on $\mathcal{A}$, corresponds with the minimal projectors $Q_i \in \mathcal{B}$ ($i = 1, \ldots, d \equiv \dim \mathcal{B}$) to a *decomposition* $\omega = \sum_{i=1}^d \omega_i$ into positive linear (hence *continuous*) functionals ω_i on $\mathcal{A}$, uniquely determined by: $P(A) = \sum_{j=1}^d \omega_j(A/\omega_j(1))Q_j \; \forall A \in \mathcal{A}$, such that $\omega_i(1) = \mu(Q_i)$ for $i = 1, \ldots, d$. With $Q_i \circ P(A) = \omega_i(A)/\omega_i(1)$ we get an equivalent *definition* for the entropy (iii) of $\gamma_1 \in CP_1(\mathcal{A})$:

$$H_\omega(\gamma_1) = \sup_{\{\sum_i \omega_i = \omega\}} \left[\sum_{i=1}^d \omega_i(1) S(\omega \circ \gamma_1 | \frac{\omega_i \circ \gamma_1}{\omega_i(1)}) \right],$$

where the supremum is taken over all finite decompositions of ω into positive linear functionals ω_i on $\mathcal{A}$ (as above). In particular, for the c.p.u. *inclusion* $i_{\mathcal{A}_1}$ of a finite-dimensional unital *sub*-*-algebra $\mathcal{A}_1 \subset \mathcal{A}$, this expression for $H_\omega(i_{\mathcal{A}_1}) \equiv H_\omega(\mathcal{A}_1)$ coincides with that given by Narnhofer and Thirring [16] in a "side-remark"; and for $\gamma_1 \in CP_1(\mathcal{A})$ where $\gamma_1 : \mathcal{A}_1 \rightarrow \mathcal{A}$ with *Abelian* C^*-algebras $\mathcal{A}_1$ resp. $\mathcal{A}$, we shall explicitly *calculate* $H_\omega(\gamma_1)$ in the "appendix".

3. For $n > 1$ as in (i) and (ii), the minimal projectors in $\mathcal{B} \equiv \bigvee_{k=1}^n \mathcal{B}_k$ are given by $Q_{(i_1,\ldots,i_n)} = Q_{i_1}^1 \cdot \ldots \cdot Q_{i_n}^n$ for $(i_1, \ldots, i_n) \equiv I_n \in \mathbf{N}^n$, and the Abelian model corresponds to a uniquely determined *multi*-index decomposition $\omega = \sum_{I_n} \omega_{I_n}$ with $\omega_{I_n}(1) = \mu(Q_{I_n})$. We denote the n single-index partial sums by

$$\omega_{i_k}^k = \sum_{I_n, i_k \text{ fixed}} \omega_{I_n},$$

such that $\omega_{i_k}^k(1) = \mu_k(Q_{i_k}^k)$ and $Q_{i_k}^k \circ E_k \circ P(A) = \omega_{i_k}^k(A)/\omega_{i_k}^k(1) \equiv \hat{\omega}_{i_k}^k(A)$. For the finite-dimensional Abelian algebra $\mathcal{B}$, the definition of the entropy S as in (iii) coincides with Def. (2.2,i) by $S(\mu) = H_\mu(\mathcal{B})$ resp. $S(\mu_k) = H_{\mu_k}(\mathcal{B}_k)$, and we get for the entropy of $\gamma_1, \ldots, \gamma_n \in CP_1(\mathcal{A})$:

$$H_\omega(\gamma_1, \ldots, \gamma_n) = \sup_{\{\sum_{I_n} \omega_{I_n} = \omega\}} [\sum_{I_n} \eta(\omega_{I_n}(\mathbf{1})) - \sum_{k=1}^n \sum_{i_k} \eta(\omega_{i_k}^k(\mathbf{1})) +$$

$$+ \sum_{k=1}^n \sum_{i_k} \omega_{i_k}^k(\mathbf{1}) S(\omega \circ \gamma_k | \hat{\omega}_{i_k}^k \circ \gamma_k)],$$

where the supremum is taken over all finite multi-index decompositions of ω (as above). In particular, for the c.p.u. *inclusions* $i_{\mathcal{A}_k}$ of *sub*algebras $\mathcal{A}_k \subset \mathcal{A}$ $(k = 1, \ldots, n)$, this expression for $H_\omega(i_{\mathcal{A}_1}, \ldots, i_{\mathcal{A}_n}) \equiv H_\omega(\mathcal{A}_1, \ldots, \mathcal{A}_n)$ coincides with that given by Connes [3]; and for an *Abelian* W^*-algebra $\mathcal{A}$ this expression again coincides with the classical expression (2.2,i): $H_\omega(\mathcal{A}_1, \ldots, \mathcal{A}_n) = H_\omega(\mathcal{A}_1 \vee \ldots \vee \mathcal{A}_n)$, cf. [4].

Definition (3.3): Let $\mathcal{A} \ni \mathbf{1}$ be a C^*-algebra with a *-automorphism $\theta \in \mathrm{Aut}(\mathcal{A})$ and with an invariant state $\omega = \omega \circ \theta$ on $\mathcal{A}$.

(i) The *entropy* of θ w.r.t. $\gamma \in CP_1(\mathcal{A})$ is defined by

$$h_\omega(\theta, \gamma) = \lim_{n \to \infty} \frac{1}{n} H_\omega(\gamma, \theta \circ \gamma, \ldots, \theta^{n-1} \circ \gamma)$$

with Def. (3.1,iii) for the r.h.s.

(ii) The *Connes-Narnhofer-Thirring* (CNT) *entropy* of θ is defined by

$$h_\omega(\theta) = \sup_{\gamma \in CP_1} h_\omega(\theta, \gamma).$$

By (3.2,1) it is sufficient to take the supremum over all c.p.u. maps $\gamma : M_d(\mathbf{C}) \to \mathcal{A} \; \forall \, d \in \mathbf{N}$.

Again, the basic property of the CNT-entropy is its conjugacy invariance $h_\omega(\theta) = h_{\omega \circ \sigma}(\sigma^{-1} \circ \theta \circ \sigma)$ for $\sigma \in \mathrm{Aut}(\mathcal{A})$; but the most *important* property is the following analogue of a *weak* version of the "Kolmogorov-Sinai" theorem for the measure-theoretic entropy (cf. [23], p. 99 and [5], p. 65):

Theorem (3.4): Let $\mathcal{A} \ni \mathbf{1}$ be a *nuclear* C^*-algebra, i.e. the *C^*-algebraic tensor product* $\mathcal{A} \otimes \mathcal{B}$ with any C^*-algebra $\mathcal{B}$ is uniquely determined (cf. [7,22]).

(i) $\mathcal{A}$ has the following "approximation property": There exist c.p.u. continuous linear maps $\sigma_n : \mathcal{A} \to \mathcal{A}_n$, $\tau_n : \mathcal{A}_n \to \mathcal{A}$ with finite-dimensional C^*-algebras $\mathcal{A}_n$ ($\forall n \in \mathbf{N}$) such that $\lim_{n \to \infty} \|\tau_n \circ \sigma_n(A) - A\| = 0$, $\forall A \in \mathcal{A}$ (Choi and Effros [2], cf. also [7]).

(ii) The CNT entropy of $\theta \in \mathrm{Aut}(\mathcal{A})$ with invariant state $\omega = \omega \circ \theta$ is given by $h_\omega(\theta) = \lim_{n\to\infty} h_\omega(\theta, \tau_n)$, with $\tau_n \in CP_1(\mathcal{A})$ as in (i); [4].

As a corollary of (ii), again the *scaling* property $h_\omega(\theta^k) = |k| \cdot h_\omega(\theta)\ \forall\, k \in \mathbf{Z}$ of the CNT entropy (originally proved for *von Neumann* algebras $\mathcal{A}$ in [4]) was proved also for *nuclear* C^*-algebras $\mathcal{A}$ in [11], which was no surprise because of the following result of [4]:

Theorem (3.5): Let $\mathcal{A} \ni \mathbf{1}$ be a nuclear C^*-algebra with $\theta \in \mathrm{Aut}(\mathcal{A})$ and with state $\omega = \omega \circ \theta$ on $\mathcal{A}$, and we denote by $\bar{\theta}$ resp. $\bar{\omega}$ the natural extensions of θ resp. ω to the von Neumann algebra $\mathcal{M}_\omega = \pi_\omega(\mathcal{A})''$ generated by the GNS-representation $\pi_\omega(\mathcal{A})$ of $\mathcal{A}$ constructed with ω (cf. [22], e.g.). Then the CNT entropy of $\bar{\theta} \in \mathrm{Aut}(\mathcal{M}_\omega)$ with invariant state $\bar{\omega} = \bar{\omega} \circ \bar{\theta}$ coincides with the original one: $h_{\bar{\omega}}(\bar{\theta}) = h_\omega(\theta)$.

In spite of this "representation-independence", the CNT entropy $h_\omega(\theta)$ of course *depends* on the θ-invariant state ω chosen on $\mathcal{A}$:

Definition (3.6): Let $\mathcal{A} \ni \mathbf{1}$ be a nuclear C^*-algebra with $\theta \in \mathrm{Aut}(\mathcal{A})$, and we denote by $E^\theta_\mathcal{A} \subset \mathcal{A}^*$ the weak-*-compact, convex (nonempty) set of all θ-invariant states $\omega = \omega \circ \theta$ on $\mathcal{A}$. The *entropy map* of the C^*-dynamical system $(\mathcal{A}, \theta)$ is the map $\bar{h}_\theta : E^\theta_\mathcal{A} \to [0, \infty]$ defined by $\bar{h}_\theta(\omega) = h_\omega(\theta)\ \forall\, \omega \in E^\theta_\mathcal{A}$ with Def. (3.3,ii) for the r.h.s.

Remarks (3.7):

1. It should be remarked that in (3.5) the von Neumann algebra $\mathcal{M}_\omega$ is *not* nuclear as a C^*-algebra in general; e.g. $\mathcal{B}(\mathcal{H})$ for $\dim \mathcal{H} = \infty$ does not have the approximation property (3.4,i). By the results of Choi and Effros (cf. [2,7]), the *nuclear* C^*-algebras $\mathcal{A}$ are exactly those for which the bidual $\mathcal{A}^{**}$ (or equivalently, the envelopping von Neumann algebra in the universal representation of $\mathcal{A}$, cf. [22]) and thus also $\mathcal{M}_\omega \equiv \pi_\omega(\mathcal{A})''$ $\forall\, \omega \in E^\theta_\mathcal{A}$, $\theta \in \mathrm{Aut}(\mathcal{A})$, is *injective*. But as remarked in [4], for nuclear $\mathcal{A}$ each $\mathcal{M}_\omega$ is even generated by an ascending *sequence* of finite-dimensional (von Neumann) subalgebras $\mathcal{M}_n \subset \mathcal{M}_{n+1} \subset \mathcal{M}_\omega$ ($\forall\, n \in \mathbf{N}$) with weakly dense union $(\bigcup_{n=1}^\infty \mathcal{M}_n)'' = \mathcal{M}_\omega$, although this is not the case for a *general* injective von Neumann algebra (such as $\mathcal{A}^{**}$, cf. [8]); and the CNT entropy of $\bar{\theta} \in \mathrm{Aut}(\mathcal{M}_\omega)$ with invariant state $\bar{\omega}$ as in (3.5) is given by $h_{\bar{\omega}}(\bar{\theta}) = \lim_{n\to\infty} h_{\bar{\omega}}(\bar{\theta}, \mathcal{M}_n)$ with Def. (3.3,i) for the r.h.s. [3,4].

2. The definition (3.6) of the entropy map seems even more "tautological" than Def. (2.4), showing that the C^*-dynamical system $(\mathcal{A}, \theta)$ is "taylor-made" for defining the map $\bar{h}_\theta$, which has been generally proved in [4] to be *concave* on $E^\theta_\mathcal{A}$ and is *affine* at least for all *asymptotically Abelian*

automorphisms θ (cf. [18]). In [17] it was shown that the map $E^{\theta}_{\mathcal{A}} \ni \omega \mapsto h_{\omega}(\theta, \gamma) \in [0, \infty]$, with Def. (3.3,i) for fixed $\gamma \in CP_1(\mathcal{A})$, is weak-*-uppersemicontinuous; but because of the ***supremum*** in Def. (3.3,ii) nothing is known as yet about the possible weak-*-uppersemicontinuity of the entropy map $\bar{h}_{\theta} : E^{\theta}_{\mathcal{A}} \to [0, \infty)$ for some $\theta \in \mathrm{Aut}(\mathcal{A})$ of a general nuclear C^*-algebra $\mathcal{A}$. A sufficient condition would be the existence of a *generator* $\gamma \in CP_1(\mathcal{A})$ such that $h_{\omega}(\theta, \gamma) = h_{\omega}(\theta)\ \forall\ \omega \in E^{\theta}_{\mathcal{A}}$ (cf. [5], pp. 107, 160; cf. also [15]).

4 The Algebraic Variational Principle

Let (X, T) be a t.d.s. (as in section 2), then the Abelian C^*-algebra $\mathcal{A} = C(X)$ is *separable* (cf. e.g. [23], p. 17), and the Gelfand space (or *spectrum*) of $\mathcal{A} \ni \mathbf{1}$ is again homeomorphic to the compact metric space X (cf. [22], p. 18). Thus the *-automorphism τ of $\mathcal{A}$ induced by T (as in (2.5,2)) uniquely determines T again, and the set $M(X, T)$ is identified by (2.5,2) with the set $E^{\tau}_{\mathcal{A}} \subset \mathcal{A}^*$ as in (3.6). We call $(\mathcal{A}, \tau)$ the C^*-dynamical system *associated* to the t.d.s. (X, T) by this one-to-one correspondence. Note that the entropy map (3.6) of $(\mathcal{A}, \tau)$ is well-defined, since *any* Abelian C^*-algebra $\mathcal{B}$ is *nuclear* (as defined in (3.4); cf. [22], p. 215).

Proposition (4.1): Let (X, T) be a t.d.s. with associated C^*-dynamical system $(\mathcal{A}, \tau)$, then we have for the entropy maps (2.4) of (X, T) resp. (3.6) of $(\mathcal{A}, \tau) : h_T(\mu) = \bar{h}_{\tau}(\omega_{\mu})\ \forall\ \mu \in M(X, T)$, with the induced state $\omega_{\mu} \in E^{\tau}_{\mathcal{A}}$ as in (2.5,2).

Proof: Let $\mu \in M(X, T)$ with W^*-algebra $\mathcal{M}_{\mu} \equiv L^{\infty}(X, \mu)$ from (2.2), and we denote by $\bar{\tau}_{\mu}$ resp. $\bar{\omega}_{\mu}$ the *-automorphism resp. state induced on $\mathcal{M}_{\mu}$ by T resp. μ as in (2.2). By Def. (2.4), we have for the l.h.s. above $h_T(\mu) = h_{\bar{\omega}_{\mu}}(\bar{\tau}_{\mu})$ with Def. (2.2,iii) for the r.h.s. – The GNS representation $\pi_{\omega_{\mu}}(\mathcal{A})$ of $\mathcal{A}$, constructed with ω_{μ} as above, is realized as the multiplication representation of $C(X)$ on the Hilbert space $\mathcal{H}_{\mu} \equiv L^2(X, \mu)$, and $\mathcal{M}_{\mu} = \pi_{\omega_{\mu}}(\mathcal{A})''$ acts by multiplication as maximally Abelian von Neumann subalgebra of $\mathcal{B}(\mathcal{H}_{\mu})$ (cf. [22], p. 103). Furthermore, the maps $\bar{\tau}_{\mu}$ resp. $\bar{\omega}_{\mu}$ induced on $\mathcal{M}_{\mu} = \pi_{\omega_{\mu}}(\mathcal{A})''$ are the natural extensions of the maps τ resp. ω_{μ} induced on $\mathcal{A}$ (by T resp. μ). By (3.7,1) together with (3.2,3), $h_{\bar{\omega}_{\mu}}(\bar{\tau}_{\mu})$ from (2.2,iii) coincides with the CNT entropy (3.3,ii) for the Abelian von Neumann algebra $\mathcal{M}_{\mu}$, and for the latter (3.5) gives $h_{\bar{\omega}_{\mu}}(\bar{\tau}_{\mu}) = h_{\omega_{\mu}}(\tau) \equiv \bar{h}_{\tau}(\omega_{\mu})$, with Def. (3.6).

Proposition (4.2): Let $\mathcal{A} \ni \mathbf{1}$ be an Abelian C^*-algebra with $\theta \in \mathrm{Aut}(\mathcal{A})$ and $\omega \in E^{\theta}_{\mathcal{A}}$. We denote by $P_1(\mathcal{A})$ or simply P_1 the set of all *positive* unital linear maps from finite-dimensional unital *Abelian* C^*-algebras into $\mathcal{A}$. Then the CNT entropy (3.3,ii) is given by $h_\omega(\theta) = \sup_{\gamma\in P_1} h_\omega(\theta,\gamma)$, with Def. (3.3,i) for the r.h.s.

Proof: Since $\mathcal{A}$ is Abelian, $CP_1(\mathcal{A})$ is the set of all *positive* unital linear maps from finite-dimensional unital C^*-algebras into $\mathcal{A}$ (cf. e.g. [22], p. 194). But $\mathcal{A}$ is always nuclear (as mentioned above), and in the approximation property (3.4,i) the finite-dimensional C^*-algebras $\mathcal{A}_n$ can be chosen to be *Abelian* as $\mathcal{A}$ is (cf. [2,7]; in fact, the approximating maps $\tau_n \in P_1(\mathcal{A})$ and $\sigma_n : \mathcal{A} \to \mathcal{A}_n$ can be easily constructed explicitly), such that by (3.4,ii) we have $h_\omega(\theta) = \lim_{n\to\infty} h_\omega(\theta,\tau_n)$ with $\tau_n \in P_1(\mathcal{A})$. By the same argument as in the proof of Prop. (2.2) in [11] we can conclude $\lim_{n\to\infty} h_\omega(\theta,\tau_n) = \sup_{\gamma\in P_1} h_\omega(\theta,\gamma)$.

Using the notation of [22, p. 4], we denote by $G(\mathcal{A})^+$ the set of all *invertible* positive elements A in the C^*-algebra $\mathcal{A} \ni \mathbf{1}$, or equivalently the set of all *strictly positive* elements $A \in \mathcal{A}^+$ with $\varphi(A) > 0$ for any nonzero positive linear functional $\varphi \in \mathcal{A}^*$ (cf. [22], p. 31). For the *Abelian* C^*-algebra $\mathcal{A} = C(X)$ as above this is equivalent to $A \in \mathcal{A}^+$ being a *strictly* positive continuous function on the compact space X; and in this case we denote by $P(\mathcal{A})$ the set of all positive linear maps γ from finite-dimensional *Abelian* C^*-algebras (with unity $\mathbf{1}$) into $\mathcal{A}$ with $\gamma(\mathbf{1}) \in G(\mathcal{A})$, such that $P(\mathcal{A}) \supset P_1(\mathcal{A})$ as in (4.2).

Definition (4.3): Let $\mathcal{A} \ni \mathbf{1}$ be an Abelian C^*-algebra. For $\gamma_1, \gamma_2 \in P(\mathcal{A})$ with $\gamma_k : \mathcal{A}_k \to \mathcal{A}$, we define the *join* $\gamma_1 \vee \gamma_2 : \mathcal{A}_1 \otimes \mathcal{A}_2 \to \mathcal{A}$ by the (linear) extension of $(\gamma_1 \vee \gamma_2)(A_1 \otimes A_2) = \gamma_1(A_1) \cdot \gamma_2(A_2)\ \forall\ A_k \in \mathcal{A}_k$ $(k = 1,2)$; such that $\gamma_1 \vee \gamma_2 \in P(\mathcal{A})$, resp. $\gamma_1 \vee \gamma_2 \in P_1(\mathcal{A})$ for $\gamma_1, \gamma_2 \in P_1(\mathcal{A})$. For $\gamma \in P_1(\mathcal{A})$ with $\gamma : \mathcal{B} \to \mathcal{A}$, we call $\gamma' \in P(\mathcal{A})$ *submap* of γ if $\gamma' : \mathcal{B}' \to \mathcal{A}$ with a unital subalgebra $\mathcal{B}' \subset \mathcal{B}$ generated (as linear span) by a *subset* of the minimal projectors in $\mathcal{B}$ (with identity $\mathbf{1}' \in \mathcal{B}'$ *different* from $\mathbf{1} \in \mathcal{B}$ in general), such that $\gamma' = \gamma_{|\mathcal{B}'}$ (but still $\gamma'(\mathbf{1}') \in G(\mathcal{A})$, by assumption).

(i) A submap $\gamma' : \mathcal{B}' \to \mathcal{A}$ of $\gamma \in P_1(\mathcal{A})$ is called *minimal*, if the (linear) dimension $\dim \mathcal{B}'$ is minimal in the set of all submaps of γ. We denote by $N(\gamma) = \dim \mathcal{B}'$ this dimension for a minimal submap $\gamma' : \mathcal{B}' \to \mathcal{A}$ of $\gamma \in P_1(\mathcal{A})$. The *entropy* of $\gamma \in P_1(\mathcal{A})$ is defined by $H(\gamma) = \log N(\gamma)$.

(ii) The "entropy" of $\tau \in \mathrm{Aut}(\mathcal{A})$ w.r.t. $\gamma \in P_1(\mathcal{A})$ is defined by

$$\bar{h}(\tau,\gamma) = \lim_{n\to\infty} \frac{1}{n} H(\gamma \vee (\tau \circ \gamma) \vee \ldots \vee (\tau^{n-1} \circ \gamma)).$$

The limit exists because of Lemma (4.4) below.

(iii) The "topological" *entropy* of $\tau \in \mathrm{Aut}(\mathcal{A})$ is defined by $\bar{h}(\tau) = \sup_{\gamma\in P_1} \bar{h}(\tau,\gamma)$.

Lemma (4.4): Let $\mathcal{A} \ni \mathbf{1}$ be an Abelian C^*-algebra, then the entropy functional (4.3,i) on $P_1(\mathcal{A})$ has the properties of subadditivity (i) and invariance (ii):

(i) $0 \leq H(\gamma_1 \vee \gamma_2) \leq H(\gamma_1) + H(\gamma_2) \ \forall \gamma_1, \gamma_2 \in P_1(\mathcal{A})$,

(ii) $H(\tau \circ \gamma) = H(\gamma) \ \forall \tau \in \mathrm{Aut}(\mathcal{A}),\ \gamma \in P_1(\mathcal{A})$.

Proof:

(i) Let γ_1' resp. γ_2' be minimal submaps of γ_1 resp. γ_2, then $\gamma_1' \vee \gamma_2'$ is a submap of $\gamma_1 \vee \gamma_2$, i.e. $N(\gamma_1 \vee \gamma_2) \leq N(\gamma_1) \cdot N(\gamma_2)$.

(ii) For any C^*-algebra $\mathcal{A} \ni \mathbf{1}$, we have $\tau(G(\mathcal{A})) = G(\mathcal{A}) \ \forall \tau \in \mathrm{Aut}(\mathcal{A})$.

Theorem (4.5): Let (X,T) be a t.d.s. with associated C^*-dynamical system $(\mathcal{A},\tau)$, then we have for the (topological) entropies (2.1,iii) of (X,T) resp. (4.3,iii) of $(\mathcal{A},\tau) : h(T) = \bar{h}(\tau)$.

Proof: We prove (i) $h(T) \geq \bar{h}(\tau)$ and (ii) $h(T) \leq \bar{h}(\tau)$.

(i) Let $\gamma \in P_1(\mathcal{A})$ with $\gamma : \mathcal{B} \to \mathcal{A}$, then γ corresponds to a finite partition of unity $\mathbf{1} = \sum_i f_i$ in $C(X)$ with $f_i(x) \geq 0$ ($\forall x \in X$), given by $\gamma(Q_i) \equiv f_i$ with the minimal projectors $Q_i \in \mathcal{B}$ ($i = 1, \ldots, \dim \mathcal{B}$); and we denote by $\alpha(\gamma) \in \mathcal{O}(X)$ the finite open cover of X uniquely determined by $\alpha(\gamma) = \{A_i | i = 1, \ldots, \dim \mathcal{B}\}$, where A_i is given by the open interior of the closed $\mathrm{supp}(f_i) \subseteq X$ (cf. also the proof of (2.3)!). Then obviously by Def. (4.3,i) for a *minimal submap* γ' of $\gamma \in P_1(\mathcal{A})$, with $\gamma'(\mathbf{1}') \in G(\mathcal{A})$, $\alpha(\gamma')$ defined as above is a *minimal subcover* of $\alpha(\gamma)$ as in Def. (2.1,i). Furthermore, by Def. (2.5,2) resp. (2.2) of the induced automorphism τ together with Def. (2.1) of the *action* by T on $\mathcal{O}(X)$, we have $\alpha(\tau \circ \gamma) = T^{-1}\alpha(\gamma)$; and by Def. (4.3) resp. (2.1) of the *join* in $P(\mathcal{A})$ resp. $\mathcal{O}(X)$ we have:

$$\alpha\Big(\bigvee_{k=0}^{n-1} \tau^k \circ \gamma\Big) = \bigvee_{k=0}^{n-1} \alpha(\tau^k \circ \gamma) = \bigvee_{k=0}^{n-1} T^{-k}\alpha(\gamma) \equiv \alpha_T^n(\gamma) \quad \forall n \in \mathbf{N}.$$

Together we see from Def. (4.3) resp. (2.1): $\forall \gamma \in P_1(\mathcal{A}) \ \exists \alpha(\gamma) \in \mathcal{O}(X)$ with $\bar{h}(\tau,\gamma) = h(T, \alpha(\gamma))$; i.e. $\bar{h}(\tau) \leq h(T)$.

(ii) Conversely, let $\beta \in \mathcal{O}(X)$ be a finite open cover $\beta = \{A_1, \ldots, A_n\}$, then there exists a finite partition of unity $\mathbf{1} = \sum_{i=1}^n f_i$ in $C(X)$ with $\mathrm{supp}(f_i) = \bar{A}_i$ ($i = 1, \ldots, n$); and we define a map $\gamma_\beta \in P_1(\mathcal{A})$ with $\gamma_\beta : \mathbf{C}^n \to \mathcal{A} = C(X)$ by linear extension of $\gamma_\beta(e_i) = f_i$, where e_i denote the minimal projectors in the C^*-algebra $\mathbf{C}^n$ (with componentwise multiplication) given by $(e_i)_j = \delta_{ij}$ ($i,j = 1, \ldots, n$). Then together with the map $\alpha : P(\mathcal{A}) \to \mathcal{O}(X)$ from (i) we have $\alpha(\gamma_\beta) = \beta$; and by (i) we can conclude: $\forall\ \beta \in \mathcal{O}\ \exists\ \gamma_\beta \in P_1$ with $\bar{h}(\tau, \gamma_\beta) = h(T, \beta)$; i.e. $\bar{h}(\tau) \geq h(T)$.

Corollary (4.6): Let (X, T) be a t.d.s. with associated C^*-dynamical system $(\mathcal{A}, \tau)$, then we have for the latter with Def. (4.3) resp. (3.6) the ***algebraic variational principle***:

(i) $\bar{h}(\tau) \geq \bar{h}_\tau(\omega)\ \forall \omega \in E^\tau_{\mathcal{A}}$, and even

(ii) $\bar{h}(\tau) = \sup_{\omega \in E^\tau_{\mathcal{A}}} \bar{h}_\tau(\omega)$.

Proof: Obvious from (2.6) together with (4.1) and (4.5).

By definition (4.3), the l.h.s. is given by

$$\bar{h}(\tau) = \sup_{\gamma \in P_1} \lim_{n\to\infty} \frac{1}{n} H(\gamma \vee (\tau \circ \gamma) \vee \ldots \vee (\tau^{n-1} \circ \gamma));$$

whereas the r.h.s. of (i) is defined by (3.6) and (3.3), and together with (4.2) is given by

$$\bar{h}_\tau(\omega) = \sup_{\gamma \in P_1} \lim_{n\to\infty} \frac{1}{n} H_\omega(\gamma, \tau \circ \gamma, \ldots, \tau^{n-1} \circ \gamma)$$

with Def. (3.1,iii) resp. (3.2,3) for the r.h.s., which could thus be compared with Def. (4.3,i) for *fixed* $\gamma \in P_1(\mathcal{A})$ as argument of *both* entropy functionals. This could "in principle" lead to a *direct* proof at least of (i), as also in [20] (as mentioned in the introduction; cf. the appendix). Actually, also (4.2) still involves a comparison of the set $CP_1(\mathcal{A})$ of *all positive* unital linear maps from finite-dimensional unital C^*-algebras into $\mathcal{A}$ (as used in Def. (3.3,ii) for Abelian $\mathcal{A}$) with the "generating" subset $P_1(\mathcal{A})$, by means of the generalized Kolmogorov-Sinai theorem (3.4,ii). But we could use Def. (4.3) of the *join* $\gamma_1 \vee \gamma_2$ also for $\gamma_1, \gamma_2 \in CP_1(\mathcal{A})$ without change (cf. [22], p. 218), such that we could equivalently define also the "topological" entropy $\bar{h}(\tau)$ as a supremum over $\gamma \in CP_1(\mathcal{A})$ in (4.3,iii), and thus we would not need (4.2) for a *direct* comparison of $\bar{h}(\tau)$ with $\bar{h}_\tau(\omega)$. – To extend Def. (4.3) not only to maps $\gamma \in CP_1(\mathcal{A})$, but also for a *non-Abelian* C^*-algebra $\mathcal{A}$ with $\tau \in \mathrm{Aut}(\mathcal{A})$ (cf. also [19]), remains a further challenge for the future.

Appendix

Let X be a compact metric space with Borel measure μ, and let $\mathcal{A} \equiv C(X)$ be the associated Abelian (separable) C^*-algebra with induced state ω on $\mathcal{A} \ni \mathbf{1}$ given by $\omega(f) = \int_X f(x)d\mu(x)\ \forall\ f \in \mathcal{A}$. As in (4.2), we denote by $P_1(\mathcal{A})$ the set of all *positive* unital linear maps from finite-dimensional unital *Abelian* C^*-algebras into $\mathcal{A}$.

Theorem: Let $\gamma_1 \in P_1(\mathcal{A})$ with $\gamma_1 : \mathcal{A}_1 \to \mathcal{A}$, and we denote the images of the minimal projectors $e_i \in \mathcal{A}_1$ by $\gamma_1(e_i) \equiv f_i \in \mathcal{A}^+$ $(i = 1, \ldots, d_1 \equiv \dim \mathcal{A}_1)$, such that $\sum_{i=1}^{d_1} f_i = \mathbf{1} \in \mathcal{A}$. Then the entropy (3.2,2) of $\gamma_1 \in P_1$ is given by the following equivalent expressions:

$$H_\omega(\gamma_1) = S(\omega \circ \gamma_1) + \sum_{j=1}^{d_1} \int_X f_j(x) \log f_j(x) d\mu(x) = \sum_{j=1}^{d_1} \int_X f_j(x) \log \frac{f_j(x)}{\omega(f_j)} d\mu(x),$$

where $S(\omega \circ \gamma_1)$ denotes the entropy as in (3.1,iii).

Proof: We choose a refining sequence $\{\xi_k\}$ of finite measurable partitions $\xi_k \in \mathcal{P}(X, \mu)$ of X into 2^k sets $\xi_k = \{A^k_{i_k} | i_k = 1, \ldots, 2^k\}$ of *equal* measure $\mu(A^k_{i_k}) = 2^{-k}$ and such that $A^k_{i_k} = A^{(k+1)}_{2i_k} \cup A^{(k+1)}_{(2i_k-1)}$ $\forall\, k \in \mathbf{N}$. We define a sequence of decompositions $\omega = \sum_{i_k} \omega^k_{i_k}$ into 2^k positive linear functionals $\{\omega^k_{i_k} | i_k = 1, \ldots, 2^k\}$ on $\mathcal{A}$ by $\omega^k_{i_k}(f) = \int_{A^k_{i_k}} f(x) d\mu(x)$ $\forall\, f \in \mathcal{A}$; such that $\omega^k_{i_k}(\mathbf{1}) = 2^{-k}$ and $\omega^k_{i_k} = \omega^{(k+1)}_{2i_k} + \omega^{(k+1)}_{(2i_k-1)}$ $\forall\, k \in \mathbf{N}$. Using these decompositions in Def. (3.2,2) for $H_\omega(\gamma_1)$, we have for the second arguments of the relative entropy terms ($\forall\, k \in \mathbf{N}$):

$$\frac{\omega^k_{i_k} \circ \gamma_1}{\omega^k_{i_k}(\mathbf{1})} = 2^k \cdot \omega^k_{i_k} \circ \gamma_1 = \frac{1}{2}(\hat{\omega}^{(k+1)}_{2i_k} + \hat{\omega}^{(k+1)}_{(2i_k-1)}) \circ \gamma_1,$$

where we denoted the normalized states as in (3.2,3). This gives by the *strict convexity* of the relative entropy in the second argument (cf. [4]):

$$\sum_{i_k} \omega^k_{i_k}(\mathbf{1}) \cdot S(\omega \circ \gamma_1 | \frac{\omega^k_{i_k} \cdot \gamma_1}{\omega^k_{i_k}(\mathbf{1})}) < \sum_{i_{(k+1)}} \omega^{(k+1)}_{i_{(k+1)}}(\mathbf{1}) \cdot S(\omega \circ \gamma_1 | \hat{\omega}^{(k+1)}_{i_{(k+1)}} \circ \gamma_1),$$

such that the supremum as in (3.2,2) taken only over *this* sequence of decompositions equals the *limit* of these expressions as $k \to \infty$. But an *arbitrary* decomposition $\omega = \sum_j \omega_j$ is given by $\omega_j(f) = \omega(p_j \cdot f)$ with $p_j \in \mathcal{A}^+$, $\sum_j p_j = \mathbf{1} \in \mathcal{A}$, and can be refined to $\omega = \sum_{i_k, j} \omega^k_{i_k j}$ with $\omega^k_{i_k j}(f) \equiv \omega^k_{i_k}(p_j \cdot f)$ $\forall\, f \in \mathcal{A}_j$, such that $\forall\, \varepsilon > 0\ \exists\ N(\varepsilon)$ and $\forall\, k > N(\varepsilon)\ \exists\ \lambda^k_{i_k j} > 0$, $\sum_j \lambda^k_{i_k j} = 1$, with $\|\omega^k_{i_k j} - \lambda^k_{i_k j} \cdot \omega^k_{i_k}\| < \varepsilon\ \forall\, i_k = 1, \ldots, 2^k; \forall\, j$. Again by the strict convexity of the relative entropy, the decomposition $\{\omega^k_{i_k j}\}$ is better than $\{\omega_j\}$ for $H_\omega(\gamma_1)$ and becomes best as $k \to \infty$. As in the proof [18] of the strong continuity of $\omega \mapsto H_\omega(\gamma_1)$, we can replace $\{\omega^k_{i_k j}\}$ by $\{\lambda^k_{i_k j} \cdot \omega^k_{i_k}\}$ as $k \to \infty$; but because of the *strict* convexity of the relative entropy, $\{\lambda^k_{i_k j} \cdot \omega^k_{i_k}\}$ gives the same value for the above expression as $\{\omega^k_{i_k}\}$, and thus the limit $k \to \infty$ for this latter sequence equals already the supremum over *all* decompositions. – With the explicit expression for the relative entropy terms of states on the *Abelian* algebra $\mathcal{A}_1$ (cf. [16], also [12]) we get:

$$H_\omega(\gamma_1) = S(\omega \circ \gamma_1) + \sum_{j=1}^{d_1} \lim_{k \to \infty} \sum_{i_k} \left[\frac{\omega^k_{i_k}(f_j)}{\omega^k_{i_k}(\mathbf{1})} \log \frac{\omega^k_{i_k}(f_j)}{\omega_{i_k}(\mathbf{1})} \right] \omega^k_{i_k}(\mathbf{1}),$$

and the first expression for $H_\omega(\gamma_1)$ results from standard integration theory. The second expression is then obvious from the explicit form of $S(\omega \circ \gamma_1)$, cf. (3.2,3).

Remark: Unfortunately, this theorem cannot be extended for n maps $\gamma_1, \ldots, \gamma_n \in P_1(\mathcal{A})$ with $\gamma_k : \mathcal{A}_k \to \mathcal{A}$, for *two* (related) reasons: First, the two additional entropy sums in the expression (3.2,3) for $H_\omega(\gamma_1, \ldots, \gamma_n)$ *spoil* the joint convexity of the *relative* entropy terms. Those two sums cancel each other only for a multi-index decomposition $\omega = \sum_{I_n} \omega_{I_n}$ (with $I_n \equiv (i_1, \ldots, i_n)$) given by a multi-index *equi*partition $\xi = \{A_{(i_1,\ldots,i_n)}\}$ of X (as in the proof above) with $i_k = 1, \ldots, d$ ($\forall\, k = 1, \ldots, n$), such that $\omega_{I_n}(\mathbf{1}) = d^{-n}$ resp. $\omega^k_{i_k}(\mathbf{1}) = d^{-1}$ for the partial sums $\omega^k_{i_k}$ (*different* from $\omega^k_{i_k}$ in the proof above!) and thus $\omega_{I_n}(\mathbf{1}) = \prod_{k=1}^n \omega^k_{i_k}(\mathbf{1})$. But still then, and secondly, these partial sums $\omega^k_{i_k}$ in the remaining relative entropy terms, integrating the functions $\gamma_k(e^k_{j_k}) \equiv f^k_{j_k} \in C(X)$ on the *disconnected* sets $\bigcup_{I_n, i_k \text{ fixed}} A_{I_n}$ (with the minimal projectors $e^k_{j_k} \in \mathcal{A}_k$; $k = 1, \ldots, n$), pick up "non-local" contributions from metrically *separated* regions of X for arbitrarily large d. – Instead of an explicit expression for $H_\omega(\gamma_1, \ldots, \gamma_n)$, we can only give the following *upper bound* (cf. [4,12]):

$$H_\omega(\gamma_1, \ldots, \gamma_n) \leq H_\omega(\bigvee_{k=1}^n \gamma_k) = S(\omega \circ \bigvee_{k=1}^n \gamma_k) + \sum_{k=1}^n \sum_{j_k} \int_X f^k_{j_k}(x) \log f^k_{j_k}(x) d\mu(x)$$

with the *join* (4.3) in $P_1(\mathcal{A})$, and by the above theorem for the r.h.s.

References

[1] Adler, R.L., Konheim, A.G. and McAndrew, M.H., Trans. Amer. Math. Soc. 114 (1965), 309-319.

[2] Choi, M.-D. and Effros, E.G., Amer. J. Math. 100 (1978), 61-79.

[3] Connes, A., C.R. Acad. Sci. Paris 301 I (1985), 1-6.

[4] Connes, A., Narnhofer, H. and Thirring, W., Comm. Math. Phys. 112 (1987), 691-719; Connes, A., Narnhofer, H. and Thirring, W., in H. Mitter and L. Pittner (eds.), *Recent Developments in Mathematical Physics*, Springer, Berlin (1987), p. 102–136.

[5] Denker, M., Grillenberger, C. and Sigmund, K., *Ergodic Theory on Compact Spaces*, Springer Lecture Notes in Math. 527, Berlin (1976).

[6] Dinaburg, E.I., Soviet Math. Dokl. 11 (1970), 13-16.

[7] Effros, E.G., in R.V. Kadison (ed.), *C*-Algebras and Applications to Physics*, Springer Lecture Notes in Math. 650, Berlin (1978), p.1–40.

[8] Elliott, G.A., Math. Scand. 39 (1976), 91-101;
Elliott, G.A., Canad. Math. Bull. 21 (1978), 415-418.

[9] Goodman, T.N.T., Bull. London Math. Soc. 3 (1971), 176-180.

[10] Goodwyn, L.W., Proc. Amer. Math. Soc. 23 (1969), 679-688.

[11] Hudetz, T., Lett. Math. Phys. 16 (1988), 151-161.

[12] Hudetz, T., *Dynamical Entropy: From Classical to Quantum Theory*, thesis (in German, unpublished), Univ. Vienna (1989).

[13] Hudetz, T., *Non-Linear Entropy Functionals and a Characteristic Invariant of Symmetry Group Actions on Infinite Quantum Systems*, to appear in the proceedings of "International Conference on Selected Topics in Quantum Field Theory and Mathematical Physics", J. Niederle (ed.), World Scientific, 1989.

[14] Kolmogorov, A.N., Dokl. Akad. Nauk. SSSR 119 (1958), 861-864;
Sinai, Ya.G., Dokl. Akad. Nauk. SSSR 124 (1959), 768-771.

[15] Narnhofer, H., *Beispiele von Algebren mit gleicher dynamischer Entropie*, Vienna preprint UWThPh-1988-42, to appear in honour of A. Uhlmann, Leipzig, 1989.

[16] Narnhofer, H. and Thirring, W., Fizika 17 (1985), 257-265.

[17] Narnhofer, H. and Thirring, W., Lett. Math. Phys. 14 (1987), 89-96.

[18] Narnhofer, H. and Thirring, W., Lett. Math. Phys. 15 (1988), 261-273.

[19] Narnhofer, H. and Thirring, W., *Algebraic K-Systems*, Vienna preprint UWThPh-1989-17, to appear.

[20] Palm, G., Z. Wahrscheinlichkeitstheorie verw. Geb. 36 (1976), 27-45;
Palm, G., Soc. Math. de France, Astérisque 40 (1976), p. 159-165.

[21] Ruelle, D., *Thermodynamic Formalism*, Enc. Math. Appl. 5, Addison-Wesley, Reading (1978).

[22] Takesaki, M., *Theory of Operator Algebras I*, Springer, New York (1979).

[23] Walters, P., *An Introduction to Ergodic Theory*, Springer Graduate Texts in Math. 79, New York (1982).

AN APPLICATION OF MAJORIZATION IN THE THEORY OF DYNAMICAL SYSTEMS

Jochen Kästner [1], Christian Zylka [2]

1. INTRODUCTION

In the present paper majorization is used to solve a problem concerning dynamical systems. We call this problem the problem of accessibility and it is explained in the following. Let us consider a tripel $(\mathcal{S}, \mathcal{P}, \underline{q}_0 \in \mathcal{S})$ where $\mathcal{S}$ denotes a state space, $\mathcal{P}$ a class of processes and $\underline{q}_0$ a distinguished state. The object of investigation is the set of states that are accessible by a permitted process starting at $\underline{q}(0) = \underline{q}_0$. This set is denoted by $F(\mathcal{S}, \mathcal{P}, \underline{q}_0)$ resp. $F(\mathcal{S}, \mathcal{P})$ for short.

Of course, only a suitable choice of $\mathcal{S}$ and $\mathcal{P}$ provides a workable mathematical task. The connected $F(\mathcal{S}, \mathcal{P})$ problems turn out to be completely solvable for the pairs $(\mathcal{S}, \mathcal{P})$ chosen by us. Via studying a simple model system we will achieve far-reaching results on the accessibility by master equations. The theorems 3...6 are presented here for the first time. Their proofs use the majorization theory developped by Schur [7] , Hardy, Littlewood, and Polya [2] , and others (cf. also the monography by Marshall and Olkin [5]). In 1971 Veinott [9] introduced a generalized version, calling it "d-majorization". This concept has been developed by Uhlmann [8] and Ruch, Schranner, and Seligman [6] mainly for physical purposes.

We consider state spaces of the following type:

$$\mathcal{S}_n := \left\{ \underline{q}: \underline{q} = (q^1, q^2, \dots, q^n),\ q^i \geqslant 0,\ \sum_{i=1}^{n} q^i = 1 \right\}$$

with $n \in \mathbb{N}$ fixed. Furtermore let ST_n be the set of the n-dimensional stochastic matrices. By this we can formulate the main relation "$\underline{q} \overset{c}{\succ} \underline{p}$" (in words: "$\underline{q}$ is $\underline{c}$-majorized by $\underline{p}$") as follows:

Definition 1:

For $\underline{q}, \underline{p}, \underline{c} \in \mathcal{S}_n$, $\underline{c} \notin \partial \mathcal{S}_n$ it holds that:

$$\underline{q} \overset{c}{\succ} \underline{p} \overset{def}{\Longleftrightarrow} \exists T \in ST_n \text{ with } \underline{q} = T\,\underline{p} \text{ and } \underline{c} = T\,\underline{c}.$$

Frequently the following sets are required:

$$G(\underline{p}, \underline{c}) := \left\{ \underline{q}: \underline{q} \overset{c}{\succ} \underline{p} \right\},$$

$$\varepsilon_1(\underline{c}) := \left\{ \underline{q}: q^1/c^1 \geqslant q^2/c^2 \geqslant \dots \geqslant q^n/c^n \right\}, \text{ and}$$

$$\varphi_1(\underline{p}, \underline{c}) := G(\underline{p}, \underline{c}) \cap \varepsilon_1(\underline{c}) .$$

Furthermore for $1 \leqslant k < l \leqslant n$ and $0 \leqslant x \leqslant 1$, $x \in \mathbb{R}$ let:

1) Karl Marx University, Dept. of Mathematics, DDR-7010 Leipzig.

2) Karl Marx University, Dept. of Physics, DDR-7010 Leipzig.

$$\hat{T}(k,l,x,\underline{c}) := \left\| \begin{array}{ccccc} 1 \ddots 1 & & & & \\ & \frac{xc^l+c^k}{c^k+c^l} & & \frac{c^k(1-x)}{c^k+c^l} & \\ & & 1 \ddots 1 & & \\ & \frac{c^l(1-x)}{c^k+c^l} & & \frac{xc^k+c^l}{c^k+c^l} & \\ & & & & 1 \ddots 1 \end{array} \right\| \begin{array}{c} \\ k \\ \\ l \\ \\ \end{array}$$

By this we define

$$K(\underline{c})_{(2)} := \{\hat{T}(k,l,x,\underline{c});\ 1 \le k < l \le n,\ 0 \le x \le 1\}.$$

The following two theorems are essential for the furter considerations.

Theorem 1 (Uhlmann [8] , Ruch et al [6]):

Let $\underline{q}, \underline{p} \in \mathcal{E}_1(\underline{c})$. Then it holds that:

$$\underline{q} \succ^{\underline{c}} \underline{p} \iff \sum_{i=1}^{m} q^i \le \sum_{i=1}^{m} p^i \quad \text{for } m = 1,2,\dots,n$$

and

Theorem 2 (Zylka [11]):

Let $\underline{q} \in \varphi_1(\underline{p},\underline{c})$. Then the matrix T (cf. def. 1) can always be chosen to be of the form $T = \prod \hat{T}_r$ with $\hat{T}_r \in K(\underline{c})_{(2)}$.

Processes are changes of states in the course of time (or of another parameter, formally $[\underline{p}(t)]_{t \ge 0}$. The first class of processes we will consider is the following:

Definition 2:

$$\hat{\mathcal{P}}_{\underline{c}} := \{ [\underline{p}(t)]_{t\ge 0}: \forall t', t'' \ (0 \le t' \le t'') \ \exists s \in \mathbb{N} \text{ and } \exists \hat{T}_1,\dots,\hat{T}_s \in K(\underline{c})_{(2)} \text{ with } \underline{p}(t'') = (\prod_{r=1}^{s} \hat{T}_r)\, \underline{p}(t') \}.$$

These processes have a tangible physical meaning. We consider a closed thermodynamical system consisting of n macroscopic bodies. The bodies are numbered arbitrarily. Then c^i denotes the heat capacity of the i-th body, q^i its heat content and therefore $q^i/c^i =: t^i$ its temperature. Suitable scaling ensures that $\sum q^i = \sum c^i = 1$. Hence in this model a state $\underline{q}$ is identified with a possible distribution of the total heat to the n bodies. In order to change states we assume an "ideal metallic wire" by means of which one can connect two of these bodies - say the bodies k and l. By this a certain amount of heat will be transported from the hotter to the cooler body and their temperatures approach one another or become equal. One realizes that the matrix modelling the above procedure is precisely of the form $\hat{T}(k,l,x,\underline{c})$. The parameter x determines the extent of equalization. Hence a $\hat{\mathcal{P}}_{\underline{c}}$-process is a sequence of such connections by the wire to different pairs of bodies and $F(\mathcal{S}_n, \hat{\mathcal{P}}_{\underline{c}}, \underline{q}_0)$ is the set of all heat distributions to the n bodies accessible from the initial distribution $\underline{q}_0$.

It is convenient to represent $\mathcal{S}_n$ as (n-1)-simplex equipped with barycentric coordinates. Thus $\mathcal{S}_3$ corresponds to a regular triangle, $\mathcal{S}_4$ to a regular tetrahedron etc., whereas their heights serve as axis of coordinates. By fixing $\underline{c}$, $\mathcal{S}_n$ obtains a special structure. Figure 1 shows the situation for n = 3 with arbitrary $\underline{c}$.

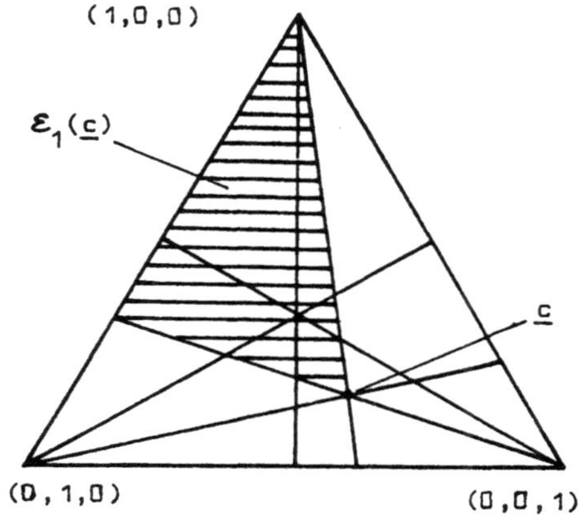

Fig. 1

2. THE SOLUTION OF THE $F(\mathcal{S}_n, \hat{\mathcal{P}}_{\underline{c}})$ PROBLEM

2.1. A DESCRIPTION OF $\varphi_1(\underline{q}_0, \underline{c})$

To begin with we assume $\underline{q}_0 \in \mathcal{E}_1(\underline{c})$. Since $\varphi_1(\underline{q}_0, \underline{c})$ is bounded and, as an intersection of two convex sets (cf. Alberti/Uhlmann [1]), itself convex, we will describe it by means of its extremal points. We will start by investigating a subclass of ST_n.

Let ϱ be a partition of the (ordered) n-tuple (1,2,...,n) written as $\varrho = (\varrho_1/\varrho_2/\dots/\varrho_k)$ with $\varrho_1 = \{1,\dots,i_1\}$, $\varrho_2 = \{i_1+1,\dots,i_2\},\dots,$ $\varrho_k = \{i_{k-1}+1,\dots,n\}$. Each partition ϱ should correspondend to a (nxn)-matrix $R(\varrho,\underline{c}) = (r_{ij})$ with

$$r_{ij} := \begin{cases} c^i / \sum\limits_{l \in \varrho_s} c^l & \text{iff } i,j \in \varrho_s \text{ for a } s \in \{1,\dots,k\} \\ 0 & \text{otherwise .} \end{cases}$$

For instance, we have for n = 3 and $\varrho = (1,2/3)$:

$$R(\varrho,\underline{c}) = \left\| \begin{matrix} \dfrac{c^1}{c^1+c^2} & \dfrac{c^1}{c^1+c^2} & 0 \\ \dfrac{c^2}{c^1+c^2} & \dfrac{c^2}{c^1+c^2} & 0 \\ 0 & 0 & 1 \end{matrix} \right\| .$$

The set of matrices constructed in this manner is denoted by $M_1(\underline{c})$ and its convex hull by conv $M_1(\underline{c})$. Using this we can describe $\varphi_1(\underline{q}_0, \underline{c})$:

<u>Theorem 3</u>:

$\underline{p} \in \varphi_1(\underline{q}_0, \underline{c})$ holds iff a matrix $A \in$ conv $M_1(\underline{c})$ with $A\,\underline{q}_0 = \underline{p}$ exists.

Proof. At first we show the sufficiency. Let $\underline{p} = A\,\underline{q}_0$ for an $A \in \operatorname{conv} M_1(\underline{c})$ and $A = R(\varrho,\underline{c})$ for some partition ϱ. Then for $1 \leq i < j \leq n$ it holds that if there exists some s with $i \in \varrho_s$ and $j \in \varrho_s$, then we have:

$$\frac{p^i}{c^i} = \frac{\sum_{l\in\varrho_s} q_0^{\,l}}{\sum_{l\in\varrho_s} c^l} = \frac{p^j}{c^j}$$

On the other hand, if $i \in \varrho_s$ and $j \in \varrho_{s'}$ with $s < s'$, then because of $\underline{q}_0 \in \mathcal{E}_1(\underline{c})$ we obtain:

$$\frac{p^i}{c^i} = \frac{\sum_{l\in\varrho_s} q_0^{\,l}}{\sum_{l\in\varrho_s} c^l} \geq \frac{q_0^{\,i_s}}{c^{i_s}} \geq \frac{q_0^{\,i_{s'-1}+1}}{c^{i_{s'-1}+1}} \geq \frac{\sum_{l\in\varrho_{s'}} q_0^{\,l}}{\sum_{l\in\varrho_{s'}} c^l} = \frac{p^j}{c^j}\ .$$

Hence it follows that $\underline{p} = A\,\underline{q}_0 \in \mathcal{E}_1(\underline{c})$. Because of the convexity of $\mathcal{E}_1(\underline{c})$ this is true for arbitrary $A \in \operatorname{conv} M_1(\underline{c})$. One can easily check that $A \in ST_n$ and $A\,\underline{c} = \underline{c}$. Hence it holds $\underline{p} \succeq^{\underline{c}} \underline{q}_0$ and therefore $p \in \varphi_1(\underline{q}_0,\underline{c})$.

Now we come to the necessity. Because of the boundedness and convexity of $\varphi_1(\underline{q}_0,\underline{c})$ it is enough to show that for each extremal point $\underline{p}$ of $\varphi_1(\underline{q}_0,\underline{c})$ there is a partition ϱ with $\underline{p} = R(\varrho,\underline{c})\,\underline{q}_0$. We will show this via contraposition.

Assume $\underline{p} \in \varphi_1(\underline{q}_0,\underline{c})$ to be not of the form $R(\varrho,\underline{c})\,\underline{q}_0$, especially $\underline{p} \neq \underline{c}$. Then $I := \{i\colon 1 \leq i \leq n,\ p^i/c^i > p^{i+1}/c^{i+1}\}$ is non-empty and we assume $I = \{i_1,\dots,i_{k-1}\}$ with $i_1 < \dots < i_{k-1}$. Now we set $\varrho = (1,\dots,i_1/i_1+1,\dots,i_2/\dots/i_{k-1}+1,\dots,n)$. Suppose it holds that $\sum_{i=1}^{j} p^i = \sum_{i=1}^{j} q_0^{\,i}$ for all $j \in I$ and by this $\sum_{i\in\varrho_s} p^i = \sum_{i\in\varrho_s} q_0^{\,i}$ for $1 \leq s \leq k$. From $p^i/c^i = p^j/c^j$ for $i,j \in \varrho_s$ then it follows $p^j = (c^j/\sum_{i\in\varrho_s} c^i)\sum_{i\in\varrho_s} q_0^{\,i}$ and therefore $\underline{p} = R(\varrho,\underline{c})\,\underline{q}_0$ which contradicts the assumption. Consequently there exists an $m \in I$ with $\sum_{i=1}^{m} p^i < \sum_{i=1}^{m} q_0^{\,i}$. Now we set

$$A := \{j\colon p^j/c^j = p^m/c^m\}, \qquad a := |A|,\ c^A := \sum_{j\in A} c^j \quad \text{and}$$

$$B := \{j\colon p^{m+1}/c^{m+1} = p^j/c^j\},\ b := |B|,\ c^B := \sum_{j\in B} c^j\ .$$

One can easily prove that $\sum_{i=1}^{j} p^i < \sum_{i=1}^{j} q_0^{\,i}$ holds for all $j \in A$ and all $j \in B$ with $j < m+b$. Moreover, we have $p^{m+b} \neq 0$. Otherwise we would obtain $p^j = 0$ for all $j > m$ and $\sum_{i=1}^{m} p^i = 1 \geq \sum_{i=1}^{m} q_0^{\,i}$ in contradiction to the choice of m. Therefore a $\mu > 0$ exists that satisfies the following conditions:

$$\mu \leq \sum_{i=1}^{j} q_0^{\,i} - \sum_{i=1}^{j} p^i \quad \text{for all } j \in A \text{ and all } j \in B \text{ with } j < m+b,$$

$$\mu \leq c^A c^B (p^{m-a}/c^{m-a} - p^{m-a+1}/c^{m-a+1}) \quad \text{iff } m-a > 0,$$

$$\mu \leq c^A c^B (p^m/c^m - p^{m+1}/c^{m+1}) \quad \text{and}$$

$$\mu \leq c^A c^B (p^{m+b}/c^{m+b} - p^{m+b+1}/c^{m+b+1}).$$

(For m+b = n, let $p^{m+b+1}/c^{m+b+1} = 0$.)

By means of this μ we define $\underline{x}$ and $\underline{y}$ by

$$x^j = p^j \quad \text{and} \quad y^j = p^j \quad \text{for } j \notin A \cup B,$$
$$x^j = p^j + \mu(c^j/c^A) \quad \text{and} \quad y^j = p^j - \mu(c^j/c^A) \quad \text{for } j \in A,$$
$$x^j = p^j - \mu(c^j/c^B) \quad \text{and} \quad y^j = p^j + \mu(c^j/c^B) \quad \text{for } j \in B.$$

Using theorem 1 one can easily check that $\underline{x}, \underline{y} \in \varphi_1(\underline{q}_0, \underline{c})$. Furthermore one realizes that $\underline{x} \neq \underline{p}$ and $\underline{p} = (\underline{x} + \underline{y})/2$. Consequently $\underline{p}$ cannot be an extremal point of $\varphi_1(\underline{q}_0, \underline{c})$, q.e.d.

For the special case $\underline{c} = (1/n, \ldots, 1/n)$ theorem 3 reproduces a well-known result by Hoffman [3]. Further generalizations of [3] will be given in [4]. Next we drop the assumption $\underline{q}_0 \in \mathcal{E}_1(\underline{c})$. By $\mathcal{E}_\pi(\underline{c})$ (for a permutation $\pi \in \mathcal{S}_n$) we will denote the cell of $\mathcal{S}_n$ that is characterized by

$$q^{\pi(1)}/c^{\pi(1)} \geq q^{\pi(2)}/c^{\pi(2)} \geq \ldots \geq q^{\pi(n)}/c^{\pi(n)}.$$

Then in analogy to theorem 3 we have for $\underline{q}_0 \in \mathcal{E}_\pi(\underline{c})$

$$\varphi_\pi(\underline{q}_0, \underline{c}) := \mathcal{E}_\pi(\underline{c}) \cap G(\underline{q}_0, \underline{c}) = \{A\ \underline{q}_0 : A \in \operatorname{conv} M_\pi(\underline{c})\}.$$

Here $M_\pi(\underline{c})$ denotes the set of matrices $R(\wp, \underline{c})$ constructed with partitions of the n-tuple $(\pi(1), \pi(2), \ldots, \pi(n))$. For instance, we have

$$\left\| \begin{matrix} \frac{c^1}{c^1 + c^3} & 0 & \frac{c^1}{c^1 + c^3} \\ 0 & 1 & 0 \\ \frac{c^3}{c^1 + c^3} & 0 & \frac{c^3}{c^1 + c^3} \end{matrix} \right\| \in M_\pi(\underline{c}) \quad \text{for } \pi = \begin{pmatrix} 1 & 2 & 3 \\ 1 & 3 & 2 \end{pmatrix}.$$

Since $\underline{q}_0$ can belong to many cells simultaneously we define more generally

$$E(\underline{q}_0, \underline{c}) := \bigcup_{\pi : \underline{q}_0 \in \mathcal{E}_\pi(\underline{c})} \mathcal{E}_\pi(\underline{c})$$

and

$$\Phi(\underline{q}_0, \underline{c}) := E(\underline{q}_0, \underline{c}) \cap G(\underline{q}_0, \underline{c}) = \bigcup_{\pi : \underline{q}_0 \in \mathcal{E}_\pi(\underline{c})} \varphi_\pi(\underline{q}_0, \underline{c}). \tag{1}$$

2.2. A CONSTRUCTING PROCEDURE FOR $F(\mathcal{S}_n, \hat{\mathcal{P}}_{\underline{c}}, \underline{q}_0)$

According to theorem 2 and definition 2 a state $\underline{p}$ lying in the same cell $\mathcal{E}_\pi(\underline{c})$ as $\underline{q}_0$ becomes accesible by a $\hat{\mathcal{P}}_{\underline{c}}$-process if and only if $\underline{p}$ is $\underline{c}$-majorized by $\underline{q}_0$. In other words we have

$$F(\mathcal{S}_n, \hat{\mathcal{P}}_{\underline{c}}, \underline{q}_0) \cap \mathcal{E}_\pi(\underline{c}) = \varphi_\pi(\underline{q}_0, \underline{c})$$

and therefore

$$F(\mathcal{S}_n, \hat{\mathcal{P}}_{\underline{c}}, \underline{q}_0) \cap E(\underline{q}_0, \underline{c}) = \Phi(\underline{q}_0, \underline{c}) . \tag{2}$$

This can be taken as the basis for our construction.

It is true that $F(\mathcal{S}_n, \hat{\mathcal{P}}_{\underline{c}}, \underline{q}_0)$ certainly contains portions not belonging to $E(\underline{q}_0, \underline{c})$, but we will show that these portions can be described analogously by means of certain "helping" states. For this purpose we set

$$\tilde{\Phi}(\underline{q}_0, \underline{c}) := \{\underline{p} \in \Phi(\underline{q}_0, \underline{c}) : \exists \pi \in \mathcal{S}_n \text{ with } \underline{p} \in \mathcal{E}_\pi(\underline{c}) \text{ and } \underline{q}_0 \notin \mathcal{E}_\pi(\underline{c})\}.$$

These are just the crossing states from $\Phi(\underline{q}_0,\underline{c})$ to the other cells.

Lemma:

It holds that $F(\mathcal{S}_n,\hat{\mathcal{P}}_{\underline{c}},\underline{q}_0) = \Phi(\underline{q}_0,\underline{c}) \cup \bigcup_{\tilde{\underline{q}}\in\tilde{\Phi}(\underline{q}_0,\underline{c})} F(\mathcal{S}_n,\hat{\mathcal{P}}_{\underline{c}},\tilde{\underline{q}})$.

Proof. Assume an accesible $\underline{p}$ with $\underline{p}\notin\Phi(\underline{q}_0,\underline{c})$, then in the course of transition from $\underline{q}_0 = \underline{p}(0)$ to $\underline{p} = \underline{p}(t)$ there will exist some t', $0 \leq t' < t$, with $\tilde{\underline{q}} := \underline{p}(t')\in\tilde{\Phi}(\underline{q}_0,\underline{c})$ and hence $\underline{p}\in F(\mathcal{S}_n,\hat{\mathcal{P}}_{\underline{c}},\tilde{\underline{q}})$. On the other hand - because of the transitivity of the accessibility we have $F(\mathcal{S}_n,\hat{\mathcal{P}}_{\underline{c}},\tilde{\underline{q}}) \subseteq F(\mathcal{S}_n,\hat{\mathcal{P}}_{\underline{c}},\underline{q}_0)$ for all $\tilde{\underline{q}}\in\tilde{\Phi}(\underline{q}_0,\underline{c})$, q.e.d.

It turns out to be sufficent to consider only a finite number of states instead of the whole set $\tilde{\Phi}(\underline{q}_0,\underline{c})$: For $\pi\in\mathcal{T}_n$ and $1 \leq k < n$ let

$\rho(\pi,k) := (\pi(1)/\pi(2)/\ldots/\pi(k),\pi(k+1)/\pi(k+2)/\ldots/\pi(n))$ and

$P_\pi(\underline{q}_0,\underline{c}) := \{R(\rho(\pi,k),\underline{c})\ \underline{q}_0 : k = 1,\ldots,n-1\}$.

Now we define

$$Q_1(\underline{q}_0,\underline{c}) := \Big(\bigcup_{\pi:\underline{q}_0\in\mathcal{E}_\pi(\underline{c})} P_\pi(\underline{q}_0,\underline{c})\Big)\cap\tilde{\Phi}(\underline{q}_0,\underline{c})$$

and call the elements of $Q_1(\underline{q}_0,\underline{c})$ daughter states of the first order of $\underline{q}_0$.

Theorem 4:

It holds that $F(\mathcal{S}_n,\hat{\mathcal{P}}_{\underline{c}},\underline{q}_0) = \Phi(\underline{q}_0,\underline{c}) \cup \bigcup_{\underline{p}\in Q_1(\underline{q}_0,\underline{c})} F(\mathcal{S}_n,\hat{\mathcal{P}}_{\underline{c}},\underline{p})$.

Proof. It is sufficent to show that for each $\tilde{\underline{q}}\in\tilde{\Phi}(\underline{q}_0,\underline{c})$ there exists a $\underline{p}\in Q_1(\underline{q}_0,\underline{c})$ with $\tilde{\underline{q}}$ and $\underline{p}$ belonging to the same cell and $\tilde{\underline{q}} \overset{c}{\succcurlyeq} \underline{p}$. From this it follows that

$$\bigcup_{\tilde{\underline{q}}\in\tilde{\Phi}(\underline{q}_0,\underline{c})} F(\mathcal{S}_n,\hat{\mathcal{P}}_{\underline{c}},\tilde{\underline{q}}) \subseteq \bigcup_{\underline{p}\in Q_1(\underline{q}_0,\underline{c})} F(\mathcal{S}_n,\hat{\mathcal{P}}_{\underline{c}},\underline{p}).$$

The opposite inclusion is simple. The lemma above completes the proof. Assume $\tilde{\underline{q}}\in\tilde{\Phi}(\underline{q}_0,\underline{c})$ and let $\mathcal{E}_\pi(\underline{c})$ be a cell that contains $\underline{q}_0$ and $\tilde{\underline{q}}$. Then there exists a k, $1 \leq k < n$, with

$$q_0^{\pi(k)}/c^{\pi(k)} > q_0^{\pi(k+1)}/c^{\pi(k+1)} \text{ and } \tilde{q}^{\pi(k)}/c^{\pi(k)} = \tilde{q}^{\pi(k+1)}/c^{\pi(k+1)}.$$

(Because of $\tilde{\Phi}(\underline{c},\underline{c}) = \emptyset$ the trivial case $\underline{q}_0 = \underline{c}$ can be excluded.)

We set $\underline{p} := R(\rho(\pi,k),\underline{c})\ \underline{q}_0$. It holds that $\underline{p}\in\mathcal{E}_\pi(\underline{c})$ and on account of

$$\frac{p^{\pi(k)}}{c^{\pi(k)}} = \frac{q_0^{\pi(k)} + q_0^{\pi(k+1)}}{c^{\pi(k)} + c^{\pi(k+1)}} = \frac{p^{\pi(k+1)}}{c^{\pi(k+1)}}$$

also $\underline{p}\in Q_1(\underline{q}_0,\underline{c})$. Additionally because of $\tilde{\underline{q}} \overset{c}{\succcurlyeq} \underline{q}_0$ we have for $m \neq k$

$$\sum_{i=1}^{m}\tilde{q}^{\pi(i)} \leq \sum_{i=1}^{m}q_0^{\pi(i)} = \sum_{i=1}^{m}p^{\pi(i)} \qquad (3)$$

If $\sum_{i=1}^{k}\tilde{q}^{\pi(i)} > \sum_{i=1}^{k}p^{\pi(i)}$ were true, then especially $\tilde{q}^{\pi(k)} > p^{\pi(k)}$ and also

$$\tilde{q}^{\pi(k+1)} = \frac{c^{\pi(k+1)}}{c^{\pi(k)}}\tilde{q}^{\pi(k)} > \frac{c^{\pi(k+1)}}{c^{\pi(k)}}p^{\pi(k)} = p^{\pi(k+1)}$$

would hold, which contradicts (3). With theorem 1 $\tilde{\underline{q}} \overset{c}{\succcurlyeq} \underline{p}$ follows, q.e.d.

This idea is pursued inductively. For each $\underline{p}\in Q_1(\underline{q}_0,\underline{c})$ one constructs $\Phi(\underline{p},\underline{c})$ and the daughter states of $\underline{p}$.

We set

$$Q_2(\underline{q}_0,\underline{c}) := \bigcup_{\underline{p}\in Q_1(\underline{q}_0,\underline{c})} Q_1(\underline{p},\underline{c})$$

and call the elements of $Q_2(\underline{q}_0,\underline{c})$ daughter states of the second order of $\underline{q}_0$, and so on. With $Q_0(\underline{q}_0,\underline{c}) := \{\underline{q}_0\}$ we arrive at the following theorem that provides the complete instruction in order to construct the sets we are searching for.

Theorem 5:

It holds that $F(\mathscr{S}_n,\hat{\mathscr{P}}_{\underline{c}},\underline{q}_0) = \bigcup_{i=0}^{n!-1} \bigcup_{\underline{p}\in Q_i(\underline{q}_0,\underline{c})} \Phi(\underline{p},\underline{c})$.

Proof. From the above we have only

$$F(\mathscr{S}_n,\hat{\mathscr{P}}_{\underline{c}},\underline{q}_0) = \bigcup_{i=0}^{\infty} \bigcup_{\underline{p}\in Q_i(\underline{q}_0,\underline{c})} \Phi(\underline{p},\underline{c}) .$$

What remains to be shown is the finiteness of the procedure. For this purpose we consider a sequence of daughter points $\underline{q}_i$ with $\underline{q}_{i+1}\in Q_1(\underline{q}_i,\underline{c})$ for all $i\geq 0$. Since $\mathscr{S}_n$ consists only of n! cells, there exists a $k\leq n!$ in such a way that for every π with $\underline{q}_k\in\mathcal{E}_\pi(\underline{c})$ there is a daughter state $\hat{\underline{q}}_\pi$ of a lower order with $\underline{q}_k\in\varphi_\pi(\hat{\underline{q}}_\pi,\underline{c})$ (e.g. $\hat{\underline{q}}_\pi=\underline{q}_j$ for some $j<k$). From this it follows that

$$\Phi(\underline{q}_k,\underline{c}) \subseteq \bigcup_{i=0}^{n!-1} \bigcup_{\underline{p}\in Q_i(\underline{q}_0,\underline{c})} \Phi(\underline{p},\underline{c}) .$$

Now we will demonstrate that this property will be bequeathed to all daughter states of $\underline{q}_k$. Let $\underline{p} = M\,\underline{q}_k$, $M\in\bigcup_{\pi\in\mathcal{T}_n} M_\pi(\underline{c})$, be a daughter state of the first order of $\underline{q}_k$ and $\underline{p}\in\mathcal{E}_\tau(\underline{c})$. If $\underline{q}_k\in\mathcal{E}_\tau(\underline{c})$ holds, then $\underline{p}\in\varphi_\tau(\hat{\underline{q}}_\tau,\underline{c})$. If $\underline{q}_k\notin\mathcal{E}_\tau(\underline{c})$ holds, then there will be a $\sigma\in\mathcal{T}_n$, $\sigma\neq\tau$, in such a way that M affects the transition from $\mathcal{E}_\sigma(\underline{c})$ to $\mathcal{E}_\tau(\underline{c})$, i.e. $M\underline{q}\in\mathcal{E}_\tau(\underline{c})$ for all $\underline{q}\in\mathcal{E}_\sigma(\underline{c})$. Hence one has $\underline{q}_k\in\mathcal{E}_\sigma(\underline{c})$ and $M\,\hat{\underline{q}}_\sigma\in\mathcal{E}_\tau(\underline{c})$. In analogy to the proof of theorem 4 $\underline{p} \overset{\underline{c}}{\succ} M\,\hat{\underline{q}}_\sigma$ follows. This argument can be applied inductively to the daughter states of higher orders of $\underline{q}_k$. Therefore we arrive at

$$\Phi(\underline{q},\underline{c}) \subseteq \bigcup_{i=0}^{n!-1} \bigcup_{\underline{p}\in Q_i(\underline{q}_0,\underline{c})} \Phi(\underline{p},\underline{c})$$

for all $\underline{q}\in Q_j(\underline{q}_0,\underline{c})$ with $j\geq n!$, q.e.d.

In figures 2 and 3 this construction is performed for

$$\underline{q}_0 = \left(\frac{14}{20}, \frac{5}{20}, \frac{1}{20}\right) \text{ and } \underline{c} = \left(\frac{4}{20}, \frac{6}{20}, \frac{10}{20}\right) .$$

Further figures and also some properties of $F(\mathscr{S}_n,\hat{\mathscr{P}}_{\underline{c}},\underline{q}_0)$ can be found in [12].

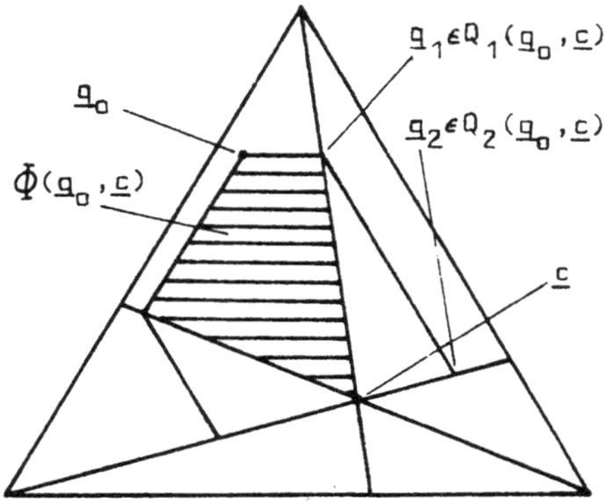

Fig. 2

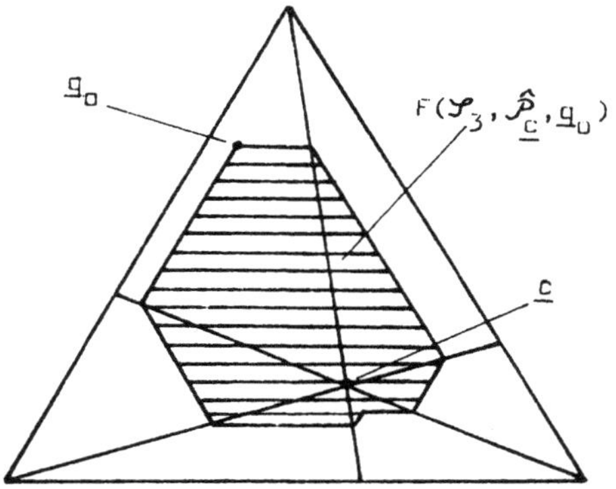

Fig. 3

3. THE SOLUTION TO A MORE GENERAL PROBLEM

Now we show that by knowing the sets $F(\mathcal{S}_n,\hat{\mathcal{P}}_{\underline{c}},\underline{q}_0)$ we already possess the solution to a more general problem. We begin with the definition of a further class of processes. For this purpose we equip $\mathcal{S}_n$ with the topology induced by $\mathbb{R}^n$.

Definition 3:

$$\mathcal{P}_{\underline{c}} := \left\{ [\underline{p}(t)]_{t\geq 0} : \underline{p}(t'') \succeq^{\underline{c}} \underline{p}(t') \text{ for all } t',t'' \text{ with } 0 \leq t' \leq t'' \text{ and } \lim_{t'\to t''} \underline{p}(t') = \underline{p}(t'') \text{ for all } t'' \geq 0 \right\}$$

Although there are certainly elements of $\mathcal{P}_{\underline{c}}$ that do not belong to $\hat{\mathcal{P}}_{\underline{c}}$ we have

Theorem 6:

For each $\underline{q}_0 \in \mathcal{S}_n$ it holds that $F(\mathcal{S}_n,\mathcal{P}_{\underline{c}},\underline{q}_0) = F(\mathcal{S}_n,\hat{\mathcal{P}}_{\underline{c}},\underline{q}_0)$.

Proof. At first we show that from $[\underline{q}(t)]_{t\geq 0} \in \mathcal{P}_{\underline{c}}$ it follows $\underline{q}(t) \in F(\mathcal{S}_n,\hat{\mathcal{P}}_{\underline{c}},\underline{q}(0))$ for all $t \geq 0$. The assumption provides $\underline{q}(t) \in G(\underline{q}(0),\underline{c})$ for all $t \geq 0$. $E(\underline{q},\underline{c})$ is (as a finite union of closed sets) a closed neighbourhood for $\underline{q}$. Hence because of the continuity of the process there exists a $\bar{t} > 0$ in such a manner that for all t with $0 \leq t \leq \bar{t}$ one has: $\underline{q}(t) \in E(\underline{q}(0),\underline{c})$ and consequently $\underline{q}(t) \in G(\underline{q}(0),\underline{c}) \cap E(\underline{q}(0),\underline{c})$. According to (1) and (2) this means $\underline{q}(t) \in F(\mathcal{S}_n,\hat{\mathcal{P}}_{\underline{c}},\underline{q}(0))$ for all t with $0 \leq t \leq \bar{t}$. Applying the latter instead of $\underline{q}(0)$ to an arbitrary $\underline{q}(t') \in F(\mathcal{S}_n,\hat{\mathcal{P}}_{\underline{c}},\underline{q}(0))$ one deduces that there exists some $t'' > t'$ with

$$\underline{q}(t) \in F(\mathcal{S}_n,\hat{\mathcal{P}}_{\underline{c}},\underline{q}(t')) \subseteq F(\mathcal{S}_n,\hat{\mathcal{P}}_{\underline{c}},\underline{q}(0)) \text{ for all } t \text{ with } t' \leq t \leq t''. \quad (4)$$

Now we suppose there is some $t^* > 0$ with $\underline{q}(t^*) \notin F(\mathcal{S}_n,\hat{\mathcal{P}}_{\underline{c}},\underline{q}(0))$. Because of the continuity of $\underline{q}(t)$ and the closed nature of $F(\mathcal{S}_n,\hat{\mathcal{P}}_{\underline{c}},\underline{q}(0))$ there must be some $t' < t^*$ with $\underline{q}(t') \in F(\mathcal{S}_n,\hat{\mathcal{P}}_{\underline{c}},\underline{q}(0))$ but $\underline{q}(t) \notin F(\mathcal{S}_n,\hat{\mathcal{P}}_{\underline{c}},\underline{q}(0))$ for all t with $t' < t \leq t^*$. But this is a contradiction to (4).

Therefore $\underline{q}(t) \in F(\mathcal{S}_n, \hat{\mathcal{P}}_{\underline{c}}, \underline{q}(0))$ is valid for all $t \geq 0$.
On the other hand it is obvious that each $\underline{q} \in F(\mathcal{S}_n, \hat{\mathcal{P}}_{\underline{c}}, \underline{q}(0))$ is accesible by a $\mathcal{P}_{\underline{c}}$-process, q.e.d.

Theorem 6 generalizes a result obtained in [10].

Equations of the form

$$\frac{dp^i}{dt} = L\, p^i, \quad i = 1,2,\dots,n, \quad \underline{p}(0) = \underline{p}_0 , \tag{5}$$

with some fixed stochastic generator L are called master equations (in a narrower sense). For $L\,\underline{c} = 0$ the solutions of (5) are $\mathcal{P}_{\underline{c}}$-processes (cf. Uhlmann [8]). Hence every solution of (5) (with $L\,\underline{c} = 0$) is completely contained in $F(\mathcal{S}_n, \mathcal{P}_{\underline{c}}, \underline{p}_0)$. With this in mind we finally consider a thermodynamic example.
We assume $\underline{c} = (1/3,\ 1/3,\ 1/3)$ and $q_0^{\,1} \geq q_0^{\,2} > q_0^{\,3}$. Instead of the wire in the model mentioned above we put a Carnot machine between two of the bodies. This machine works as long as the temperatures of the bodies are equal. The corresponding calculation that must be neglected here (due to reason of space) shows that the equalizing temperature is just the geometric mean of both initial temperatures. (In the case of the wire one obviously obtains the aritnmetic mean.) Hence twice the difference between the arithmetic and the geometric mean is the mechanical work gained.
It can be used to heat up another body. If the Carnot machine works between bodies 2 and 3 and body 1 is heated up we get:

$$\hat{q}^1 = q_0^{\,1} + 2\left(\frac{q_0^{\,2} + q_0^{\,3}}{2} - \sqrt{q_0^{\,2}\, q_0^{\,3}}\right)$$

$$\hat{q}^2 = \sqrt{q_0^{\,2}\, q_0^{\,3}}$$

$$\hat{q}^3 = \sqrt{q_0^{\,2}\, q_0^{\,3}}\,.$$

Because of $\hat{q}^1 > q_0^{\,1}$ it follows $\hat{\underline{q}} \notin F(\mathcal{S}_3, \mathcal{P}_{\underline{c}}, \underline{q}_0)$. Because of $\hat{q}^3 > q_0^{\,3}$ it also holds that $\underline{q}_0 \notin F(\mathcal{S}_3, \mathcal{P}_{\underline{c}}, \hat{\underline{q}})$. This means that the "Carnot - pump - process" described can neither by pure heat conduction nor by any other $\mathcal{P}_{\underline{c}}$-processes be reversed. With the above considerations of master equation accessibility we can in particular conclude that there is no master equation whose solution can connect $\hat{\underline{q}}$ and $\underline{q}_0$ (cf. figure 4).

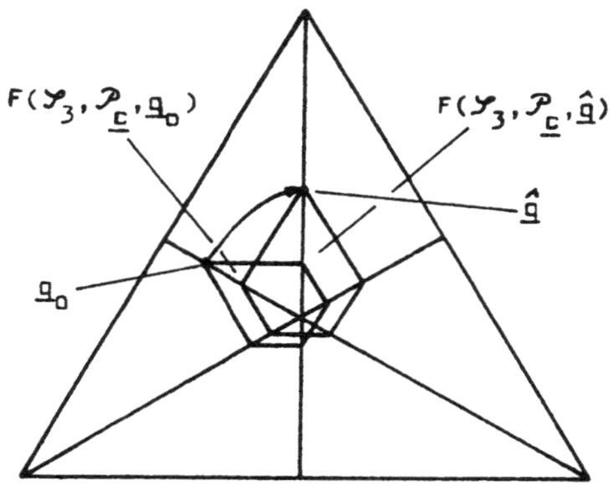

Fig. 4

References

[1] Alberti, P.M. and A. Uhlmann: Dissipative Motion in State Spaces. B.G.Teubner, Leipzig 1981.

[2] Hardy, G.H., J.E. Littlewood and G. Polya: Some simple inequalities satisfied by convex functions. Mess. Math. 58 (1929), 145-152.

[3] Hoffman, A.J.: A special class of double stochastic matrices. Aequationes Math. 2 (1969), 319-326.

[4] Kästner, J. and Ch. Zylka: On generalized uniformly tapered matrices. (to be published)

[5] Marshall, A.W. and I. Olkin: Inequalities: Theory of Majorization and Its Applications. Academic Press, New York 1979.

[6] Ruch, E., R. Schranner and T.H. Seligman: The mixing distance. J. Chem. Phys. 69 (1978), 386-392.

[7] Schur, I.: Über eine Klasse von Mittelbildungen mit Anwendungen auf die Determinantentheorie. Sitzungsber. Berl. Math. Gesellsch. 22 (1923), 9-20.

[8] Uhlmann, A.: Ordnungsstrukturen im Raum der gemischten Zustände und die H-Theoreme von Felderhof-van Kampen. Wiss. Zeitschr. Karl-Marx-Univ. Leipzig, Math.-Naturwiss. R. 27 (1978), 213-228.

[9] Veinott, A.F.: Least d-majorized network flows with inventory and statistical applications. Management Sci. 17 (1971), 547-567.

[10] Zylka, Ch.: A note on the attainability of states by equalizing processes. Theor. Chim. Acta 68 (1985), 363-377.

[11] Zylka, Ch.: Eine Verallgemeinerung des Satzes von Muirhead. Wiss. Zeitschr. Karl-Marx-Univ. Leipzig, Math.-Naturwiss. R. 34 (1985), 598-601.

[12] Zylka, Ch.: Which states are accesible by continuos c-concave processes? Publ. Joint Inst. Nucl. Res. E5-86-173, Dubna 1986.

ON APPLICATION OF LJAPUNOV FUNCTIONS AND COMPARISON SYSTEMS TO ANALYSIS OF STABILITY FOR NONLINEAR ORDINARY DIFFERENTIAL EQUATIONS

by N. K o k s c h [1)]

The aim of the paper is to investigate of nonlocal stability of complicated, in general not exactly known solutions of systems of ODE by means of the direct method of A. M. Ljapunov in connection with comparison theorems. Starting from the original system a comparison system is constructed by means of the Ljapunov function with respect to nonlocal information of the given system. The comparison system has a simpler structure than the original one, i. e. has a lower order and is of monotone type. This approach can be considered as some ***nonlocal method of reduction*** (see for example [1,2]).

As an example the Duffing equation is considered.

1. Theorem of Stability

We consider the ODE system

$$\dot{\mathbf{x}} = \mathbf{f}(t,\mathbf{x}), \tag{1}$$

where $\mathbf{f} \in C(\mathbb{R} \times \mathbb{R}^n \longrightarrow \mathbb{R}^n)$ and for any $t_0 \in \mathbb{R}$ and any $\mathbf{x}_0 \in \mathbb{R}^n$ the existence of the solution $\mathbf{x}(.)$ of (1) satisfying $\mathbf{x}(t_0) = \mathbf{x}_0$ on the interval $[t_0,\infty)$ is always assumed.

The following theorem [3] generalizes some ideas of stability theorems [4 - 10]: By application of the Ljapunov function $\mathbf{V}$ and by using some information on a special solution $\mathbf{x}_1(.)$ of (1) (f. e. a periodic one) given by some function $E_1 : \mathbb{R} \longrightarrow \mathfrak{P}(\mathbb{R}^n)$ as well as the information on all solutions given by some function $E_2 : \mathbb{R} \longrightarrow \mathfrak{P}(\mathbb{R}^n)$ a simpler comparison system is assigned to the system (1), where the stability of the stationary solution of the comparison system implies the stability of the in general not exactly known solution $\mathbf{x}_1(.)$.

---

1) Dresden University of Technology, Department of Mathematics, Mommsenstr. 13, Dresden, GDR - 8027

Theorem : *Let* $\mathbf{x}_1(.) : [t_\alpha,+\infty) \rightarrow \mathbb{R}^n$ *be a solution of the system* (1). *Let there exist*

- *a number* $m \in \mathbb{N}$,
- *functions* $E_i : \mathbb{R} \rightarrow \mathfrak{P}(\mathbb{R}^n)$, $i=1,2$, *such that* $\mathbf{x}_1(t) \in E_1(t)$, $E_1(t) \subseteq E_2(t)$ *and* $\overline{U_{\varepsilon_0}(\mathbf{x}_1(t))} \subseteq E_2(t)$ *with some* $\varepsilon_0 > 0$ *for any* $t \geq t_\alpha$. *Moreover, for all solutions* $\mathbf{x}_2(.)$ *of* (1) *there exists some time* $t_0 \geq t_\alpha$ *such that* $\mathbf{x}_2(t) \in E_2(t)$ *whenever* $t \geq t_0$,
- *a Ljapunov function* $\mathbf{V} \in C(\mathbb{R} \times D \rightarrow \mathbb{R}^m)$ *with* $\mathbf{V}(t,\mathbf{0}) = 0$ *for all* $t \in \mathbb{R}$ *and* $\max_j V_j(t,\mathbf{d}) \geq w(|\mathbf{d}|) > 0$ *for all* $t \in \mathbb{R}$ *and* $\mathbf{d} \in D \setminus \{\mathbf{0}\}$, *where* w *is some strictly monotone increasing continuous function and* $D \subseteq \mathbb{R}^n$ *is a domain satisfying* $D \supseteq \{ \mathbf{x}_2-\mathbf{x}_1 \mid \exists\, t : \mathbf{x}_1 \in E_1(t),\ \mathbf{x}_2 \in E_2(t) \}$,
- *a continuous and with respect to second argument quasimonotone increasing function* $\mathbf{F} : \Omega \rightarrow \mathbb{R}^m$ *with* $\mathbf{F}(t,\mathbf{0}) = \mathbf{0}$ *for all* $t \in \mathbb{R}$, *where* $\Omega \subseteq \mathbb{R} \times \mathbb{R}^m$ *with* $\Omega = \{ (t,\Phi(t)) \mid t \in \mathbb{R} \}$ *and* $\mathbf{V}(t,\mathbf{x}_2-\mathbf{x}_1) \in \Phi(t)$ *for all* $\mathbf{x}_i \in E_i(t)$, $i = 1,2$, *and* $t \in \mathbb{R}$,

such that the following conditions are satisfied:

(*a*) *For all* $\mathbf{x}_1 \in E_1(t)$, $\mathbf{x}_2 \in E_2(t)$ *and all* $t \geq t_\alpha$

$$\dot{\mathbf{V}}(t,\mathbf{x}_2-\mathbf{x}_1,\mathbf{x}_1) \leq \mathbf{F}(t,\mathbf{V}(t,\mathbf{x}_2-\mathbf{x}_1)) \tag{2}$$

holds, where $\dot{\mathbf{V}}(t,\mathbf{d},\mathbf{x}) = \frac{\partial}{\partial t}\mathbf{V}(t,\mathbf{d}) + \frac{\partial}{\partial \mathbf{d}}\mathbf{V}(t,\mathbf{d})\big(\mathbf{f}(t,\mathbf{x}+\mathbf{d})-\mathbf{f}(t,\mathbf{x})\big)$ *denotes the derivative of* $\mathbf{V}(t,\mathbf{d})$ *with respect to* (1) *for* $\mathbf{x} \in \mathbb{R}^n$.

(*b*) *The solution* $\mathbf{u} \equiv \mathbf{0}$ *of the ODE system of order* m *on* Ω

$$\dot{\mathbf{u}} = \mathbf{F}(t,\mathbf{u}) \tag{3}$$

is global asymptotically stable.

Then the solution $\mathbf{x}_1(.)$ *of* (1) *is global asymptotically stable.*

The proof of the theorem is based on a generalization of the Tschaplygin inequality (Theorem of Ważewski) [4, 9 - 11].

Our theorem contains weaker conditions than the known results and, therefore, allows also to study the global asymptotically stability of nonstationary solutions on a larger set of parameters at least by means of a numerical investigation of (3).

For an example we refer to the example in section 2, for which better results are obtained although the same Ljapunov function as in [12] is used.

Remark 1:
Condition (*b*) can be replaced by another one which turns out to be simpler to check however it requires a more complicated description. For it let be $\overline{\mathbf{V}} = (\overline{V}_1,\ldots,\overline{V}_m)^T$ with $\overline{V}_j : \mathbb{R} \rightarrow \mathbb{R} \cup \{+\infty\}$ and

$$\bar{V}_j(t) \geq \sup_{x_1 \in E_1(t),\; x_2 \in E_2(t)} V_j(t, x_2 - x_1) \tag{4}$$

for $j = 1, \ldots, m$.

The new condition reads now:

(*b'*) *For all* $t_0 > t_\alpha$ *there exist a* $t_1 \geq t_0$, *such that* $|\bar{\mathbf{V}}(t_1)| < \infty$ *and for the upper solution* $\mathbf{u}(.) : [t_1 - r, +\infty) \longrightarrow \mathbb{R}^m$ *of* (3) *with* $\mathbf{u}(t_1) = \bar{\mathbf{V}}(t_1)$ *there holds* $\lim_{t \to \infty} \max_j u_j(t) = 0$.

Obviously, condition (*b*) and the existence of a function $\bar{\mathbf{V}}$ with (4) such that for any $t_0 > t_\alpha$ there exists a $t_1 > t_0$ with $|\bar{\mathbf{V}}(t_1)| \leq \infty$ imply the condition (*b'*).

This simpler condition allows a numerical utilization of the criterion, after a basin of attraction of the stationary solution $\mathbf{u} \equiv \mathbf{0}$ is determined by approximate methods of stability theory. It has to be proved only that the extremal solutions mentioned in (*b'*) reach the basin of attraction during a finite time of integration.

If the comparison system with respect to time is periodic, then we have to investigate only one extremal solution on a bounded time interval.

Remark 2:
Let be $M(t) = \{ \mathbf{v} \in \mathbb{R}^m \mid \exists\, \mathbf{x}_1 \in E_1(t),\ \mathbf{x}_2 \in E_2(t) : \mathbf{v} = \mathbf{V}(t, \mathbf{x}_2 - \mathbf{x}_1) \}$ for $t \in \mathbb{R}$. Then any function $\mathbf{H} = (H_1, \ldots, H_m)^T$ with $H_j : \Omega \longrightarrow \mathbb{R} \cup \{+\infty\}$ and

$$H_j(t, \varphi) = \sup_{\substack{\mathbf{x}_1 \in E_1(t),\; \mathbf{x}_2 \in E_2(t) \\ V_j(t, \mathbf{x}_2 - \mathbf{x}_1) = \varphi_j \\ V_k(t, \mathbf{x}_2 - \mathbf{x}_1) \leq \varphi_k}} \dot{V}_j(t, \mathbf{x}_2 - \mathbf{x}_1, \mathbf{x}_1)$$

for $j = 1, \ldots, m$, $t \in \mathbb{R}$, $\varphi \in M(t)$ is quasimonotone increasing with respect to second argument and satisfies the condition (a) of the theorem with $\mathbf{F} = \mathbf{H}$. For any other quasimonotone increasing function $\mathbf{G} = (G_1, \ldots, G_m)^T$ with $G_j : \Omega \longrightarrow \mathbb{R} \cup \{\infty\}$, satisfying the condition (a), where $\mathbf{F}$ is replaced by $\mathbf{G}$, the inequality $\mathbf{H}(t, \varphi) \leq \mathbf{G}(t, \varphi)$ holds for all $t \in \mathbb{R}$ and all $\varphi \in M(t)$.

This will give us a hint for the construction of the function $\mathbf{F}$: $\mathbf{F}$ should be a continuous, quasimonotone increasing majorant of $\mathbf{H}$ possibly closed to $\mathbf{H}$.

2. Application to the investigation of stability of the Duffing-System

We consider the Duffing equation [12 - 16]:

$$\ddot{x}(t) + c\cdot\dot{x}(t) + a\cdot x(t) + b\cdot x(t)^3 = e(t) \tag{5}$$

with continuous T-periodic excitation e and nonnegative coefficients a, b, c with $4\cdot a > c^2$. It is a special case of the Liénard equation

$$\ddot{x}(t) + f(x(t))\cdot\dot{x}(t) + g(x(t)) = e(t),$$

such that the following results may be also obtained for more general equations.

With $x_1 = x$ and $x_2 = \dot{x}$ we get the system

$$\begin{aligned} \dot{x}_1 &= x_2 \\ \dot{x}_2 &= -a\cdot x_1 - b\cdot x_1^3 - c\cdot x_2 + e \ . \end{aligned} \tag{6}$$

The equation (5) with $x_1 = x$, $E(t) = \int_0^t e(\tau)\, d\tau$ and $\mathbf{f}(t,\mathbf{x}) = \begin{pmatrix} x_2 - c\cdot x_1 + E(t) \\ -a\cdot x_1 - b\cdot x_1^3 \end{pmatrix}$ is also equivalent to the system

$$\dot{\mathbf{x}} = \mathbf{f}(t,\mathbf{x}). \tag{7}$$

Let be $M \geq |e(t)|$.

N. P. Bhatia has used system (7) for the investigation of stability of the T-periodic solution of (5). For it he has determined an attractor localization for the attractor of system (6).

In what follows we describe, how by means of already known Ljapunov functions as well as by our theorem and by a numerical evaluation we can get statements on the global asymptotic stability of the periodic solution of (5). First we determine comparison systems for the calculation of an enclosing E_1 of a T-periodic solution $\bar{\mathbf{x}}(.)$ of (7), for the calculation of an attractor localization E_2 of (7) and for the investigation of stability of $\bar{\mathbf{x}}(.)$. Then we study the local stability of the comparison system (3) and determine a basin of attraction for (3). Using E_2 we get initial values $\bar{V}(t)$ for application of remark 1 and get also an estimate of the Hausdorff dimension of the attractor of (5).

Choice of the Ljapunov functions and construction of the comparison systems

For $\mathbf{z} \in \mathbb{R}^2$, $\mathbf{d} \in \mathbb{R}^2$ and $t \in \mathbb{R}$ we define the function

$$W(\mathbf{z},\mathbf{d}) = (\alpha(\mathbf{z})\cdot d_1^2 - c\cdot d_1 d_2 + d_2^2)/2 + b(z_1^2 + z_1 d_1 + d_1^2/4)\cdot d_1^2,$$

where $\alpha(\mathbf{z}) = a + bz_1^2$.

With $A(\mathbf{z}) = \alpha + 3bz_1^2$, $\mathbf{H}_1(\mathbf{z}) = \frac{1}{2}\begin{pmatrix} \alpha(\mathbf{z}) & -c/2 \\ -c/2 & 1 \end{pmatrix}$, $\mathbf{H}_2(\mathbf{z}) = \frac{1}{2}\begin{pmatrix} A(\mathbf{z}) & -c/2 \\ -c/2 & 1 \end{pmatrix}$, $Q_i(\mathbf{z},\mathbf{d}) = \mathbf{d}^*\mathbf{H}_i(\mathbf{z})\mathbf{d}$, $i = 1,2$, $J_1(\mathbf{z},\mathbf{d}) = b(z_1^2+z_1d_1+d_1^2/4)\cdot d_1^2$ and $J_2(\mathbf{z},\mathbf{d}) = b(z_1d_1+d_1^2/4)\ d_1^2$ we get:

$$W(\mathbf{z},\mathbf{d}) = Q_1(\mathbf{z},\mathbf{d}) + J_1(\mathbf{z},\mathbf{d}) = Q_2(\mathbf{z},\mathbf{d}) + J_2(\mathbf{z},\mathbf{d}),$$

where $J_1(\mathbf{z},\mathbf{d}) \geq 0$ holds and $J_2(\mathbf{z},\mathbf{d})$ does not contain quadratic terms in respect to $\mathbf{d}$. If $\mathbf{z}(.)$ is a continuously differentiable T-periodic function, then by using the function $V(t,\mathbf{d}) := W(\mathbf{z}(t),\mathbf{d})$ we study the difference $\mathbf{d}(t) = \mathbf{x}(t)-\mathbf{z}(t)$ between a solution $\mathbf{x}(.)$ of (7) and this function $\mathbf{z}(.)$. If $\mathbf{z}(.) = \bar{\mathbf{x}}(.)$ is a T-periodic solution of (7) then with $W(\mathbf{z}(.),.)$ we get a stability statement, and if $\mathbf{z}(.)$ is a T-periodic approximation for a T-periodic solution $\bar{\mathbf{x}}(.)$ then with $W(\mathbf{z}(.),.)$ we construct an enclosing of $\bar{\mathbf{x}}(.)$. Therefore, $W(\mathbf{z}(.),.)$ can be understand as a Ljapunov function.

Let be $\mathbf{D}(t,\mathbf{z},\mathbf{p}) = \mathbf{f}(t,\mathbf{z}) - \mathbf{p}$. If $\mathbf{z}(.)$ is a continuously differentiable function then is $\mathbf{D}(t,\mathbf{z}(t),\dot{\mathbf{z}}(t))$ the defect of $\mathbf{z}(.)$ with respect to system (7).

We define $W'(t,\mathbf{d},\mathbf{z},\mathbf{p}) := \frac{\partial}{\partial \mathbf{z}} W(\mathbf{z},\mathbf{d})\mathbf{p} + \frac{\partial}{\partial \mathbf{d}} W(\mathbf{z},\mathbf{d})\big(\mathbf{f}(t,\mathbf{z}+\mathbf{d}) - \mathbf{p}\big)$. Then with $V(t,\mathbf{d}) := W(\mathbf{z}(t),\mathbf{d})$ for the function $\dot{V}$ from the theorem we get $\dot{V}(t,\mathbf{d},\mathbf{z}) = W'(t,\mathbf{d},\mathbf{z},\mathbf{f}(t,\mathbf{z}))$ and for any solutions $\mathbf{x}(.)$ of (7) and any continuously differentiable function $\mathbf{z}(.)$ and $\mathbf{d}(t) = \mathbf{x}(t)-\mathbf{z}(t)$ the following equations hold

$$\frac{d}{dt} V(t,\mathbf{d}(t)) = \frac{d}{dt} W(\mathbf{z}(t),\mathbf{d}(t))$$

$$\frac{d}{dt} V(t,\mathbf{d}(t)) = W'(t,\mathbf{d}(t),\mathbf{z}(t),\dot{\mathbf{z}}(t)). \tag{8}$$

Moreover, if $\mathbf{z}(.)$ is also a solution of (7) then even

$$\frac{d}{dt} V(t,\mathbf{d}(t)) = \dot{V}(t,\mathbf{d}(t),\mathbf{z}(t)). \tag{9}$$

We describe first analytical construction of the necessary comparison systems.

There holds $W'(t,\mathbf{d},\mathbf{z},\mathbf{p}) = \sum_{i=1}^{4} U_i(t,\mathbf{d},\mathbf{z},\mathbf{p})$, where

$$U_1(t,\mathbf{d},\mathbf{z},\mathbf{p}) = \big(A(\mathbf{z})d_1 - c/2\cdot d_2\big)\cdot D_1 + \big(-c/2\cdot d_1 + d_2\big)\cdot D_2,$$
$$U_2(t,\mathbf{d},\mathbf{z},\mathbf{p}) = \big(-A(\mathbf{z})c + 3bz_1(p_1+D_1)\big)\cdot d_1^2 + (cd_1d_2 - 2d_2^2)\,c/2,$$
$$U_3(t,\mathbf{d},\mathbf{z},\mathbf{p}) = \big(-1.5bcz_1 + b(p_1+D_1)\big)\cdot d_1^3,$$
$$U_4(t,\mathbf{d},\mathbf{z},\mathbf{p}) = -bc/4\cdot d_1^4$$

and $D_1 = D_1(t,\mathbf{d},\mathbf{z})$, $D_2 = D_2(t,\mathbf{d},\mathbf{z})$ are.

Let be $\gamma_i(t,\mathbf{z},\mathbf{p}) = \sup_{Q_1(\mathbf{z},\mathbf{d}) = 1} U_i(t,\mathbf{d},\mathbf{z},\mathbf{p})$. Then the inequalities

$$W'(t,\mathbf{d},\mathbf{z},\mathbf{p}) \le \gamma_1(t,\mathbf{z},\mathbf{p})\cdot\sqrt{Q_1(\mathbf{z},\mathbf{d})} + \sum_{i=2}^{4} U_i(t,\mathbf{d},\mathbf{z},\mathbf{p})$$

$$\le \gamma_1(t,\mathbf{z},\mathbf{p})\cdot\sqrt{W(\mathbf{z},\mathbf{d})} + \sum_{i=2}^{4} U_i(t,\mathbf{d},\mathbf{z},\mathbf{p})$$

hold, because of $\gamma_1(t,\mathbf{z},\mathbf{p}) \ge 0$ and $W(\mathbf{z},\mathbf{d}) \ge Q_1(\mathbf{z},\mathbf{d})$.

Let be

$$\gamma_2(t,\mathbf{z},\mathbf{p}) = \sup_{Q_2(\mathbf{z},\mathbf{d}) = 1} U_2(t,\mathbf{d},\mathbf{z},\mathbf{p}) = -c + \frac{6b\left(z_1[y_1+D_1(t,\mathbf{z},\mathbf{p})]\right)_+}{A(\mathbf{z})-c^2/4},$$ where

$z_+ = z$ when $z > 0$ and otherwise $z_+ = 0$.

We have now

$$W'(t,\mathbf{d},\mathbf{z},\mathbf{p}) \le \sum_{i=1}^{2} \gamma_i(t,\mathbf{z},\mathbf{p})\cdot\left(W(\mathbf{z},\mathbf{d})\right)^{i/2} +$$

$$+ \sum_{i=3}^{4} U_i(t,\mathbf{d},\mathbf{z},\mathbf{p}) - \gamma_2(t,\mathbf{z},\mathbf{p})\cdot[\,W(\mathbf{z},\mathbf{d}) - Q_2(\mathbf{z},\mathbf{d})\,]$$

$$\le \sum_{i=1}^{2} \gamma_i(t,\mathbf{z},\mathbf{p})\cdot\left(W(\mathbf{z},\mathbf{d})\right)^{i/2} + \sum_{i=3}^{4} U_i'(t,\mathbf{d},\mathbf{z},\mathbf{p})$$

where $\quad U_3'(t,\mathbf{d},\mathbf{z},\mathbf{p}) = U_3(t,\mathbf{d},\mathbf{z},\mathbf{p}) - \gamma_2(t,\mathbf{z},\mathbf{p})\cdot bz_1d_1^3$

and $\quad U_4'(t,\mathbf{d},\mathbf{z},\mathbf{p}) = U_4(t,\mathbf{d},\mathbf{z},\mathbf{p}) - \gamma_2(t,\mathbf{z},\mathbf{p})\cdot bd_1^4/4 \le 0.$

Since for

$$\gamma_3(t,\mathbf{z},\mathbf{p}) = b\cdot|-1.5c\cdot z_1+p_1+D_1-\gamma_2(t,\mathbf{z},\mathbf{p})z_1|\cdot\left(\frac{2}{\alpha(\mathbf{z}) - c^2/4}\right)^{3/2}$$

the inequalities

$$\gamma_3(t,\mathbf{z},\mathbf{p}) \ge \sup_{Q_1(\mathbf{z},\mathbf{d}) = 1} U_3'(t,\mathbf{d},\mathbf{z},\mathbf{p})$$

and, because of $Q_1(\mathbf{z},\mathbf{d}) \le W(\mathbf{z},\mathbf{d})$,

$$U_3'(t,\mathbf{d},\mathbf{z},\mathbf{p}) \le \gamma_3(t,\mathbf{z},\mathbf{p})\left(W(\mathbf{z},\mathbf{d})\right)^{3/2}$$

hold, we get with

$$\Gamma(t,w,\mathbf{z},\mathbf{p}) = \sum_{i=1}^{3} \gamma_i(t,\mathbf{z},\mathbf{p})\cdot w^{i/2}$$

the inequality

$$W'(t,\mathbf{d},\mathbf{z},\mathbf{p}) \le \Gamma(t,W(\mathbf{z},\mathbf{d}),\mathbf{z},\mathbf{p}). \tag{10}$$

Construction of the enclosing E_1

Let $\mathbf{z}_1(.)$ be now a continuously differentiable T-periodic approximation for a T-periodic solution $\bar{\mathbf{x}}(.)$ of (7). We consider the equation

$$\dot{v}(t) = \Gamma(t,v(t),\mathbf{z}_1(t),\dot{\mathbf{z}}_1(t)). \tag{11}$$

Let $v_1(.)$ be a solution of (11) satisfying $v_1(t) \geq v_1(t+T) > 0$ on the time interval [0,T]. We extend this function T-periodically to $(-\infty,\infty)$. According to the construction of (11) and taking in consideration the inequality (8) for any solution $\mathbf{x}(.)$ of (7) for

$$V_1(t) = W(\mathbf{z}_1(t), \mathbf{x}(t)-\mathbf{z}_1(t))$$

there holds the inequality

$$\dot{V}_1(t) \leq \Gamma(t, V_1(t), \mathbf{z}_1(t), \dot{\mathbf{z}}_1(t)).$$

Hence $V_1(t_0) \leq v_1(t_0)$ implies the inequality $V_1(t) \leq v_1(t)$ for all $t \geq t_0$. Therefore, the function $E_1 : \mathbb{R} \longrightarrow \mathfrak{P}(\mathbb{R}^2)$ defined by

$$E_1(t) = \{ \mathbf{x} \mid W(\mathbf{z}_1(t), \mathbf{x}-\mathbf{z}_1(t)) \leq v_1(t) \}$$

for any $t \in \mathbb{R}$ satisfies the conditions of the theorem.

Construction of the attractor localization E_2

For it we have to choose a continuously differentiable T-periodic function $\mathbf{z}_2(.)$ such that $\gamma_3(t,\mathbf{z}_2(t),\dot{\mathbf{z}}_2(t)) = 0$ holds for all $t \in \mathbb{R}$.

Let be $\mathbf{z}_2(t) = (0,-E(t))^T$. Then we get

$$\mathbf{D}(t,\mathbf{z}_2(t),\dot{\mathbf{z}}_2(t)) = (0,-e(t))^T,$$
$$\gamma_3(t,\mathbf{z}_2(t),\dot{\mathbf{z}}_2(t)) = 0,$$
$$\gamma_2(t,\mathbf{z}_2(t),\dot{\mathbf{z}}_2(t)) = -c$$

and

$$\gamma_1(t,\mathbf{z}_2(t),\dot{\mathbf{z}}_2(t)) = \sqrt{2}\cdot|e(t)|$$

for all $t \in \mathbb{R}$.

We consider now the equation

$$\dot{v}(t) = \Gamma(t,v(t),\mathbf{z}_2(t),\dot{\mathbf{z}}_2(t))$$

t.e.

$$\dot{v}(t) = \sqrt{2}\cdot|e(t)|\cdot\sqrt{v(t)} - c\cdot v(t). \tag{12}$$

With $w^2 = v$, $w \geq 0$ equation (12) can be transformed into a linear equation. Therefore the T-periodic solution $v_2(.)$ of (12) we get by calculation of some integrals:

$$v_2(t) = \left(w_0\cdot\exp(-ct/2)+\sqrt{2}/2\cdot\int_0^t \exp(-c(t-\tau)/2)\cdot|e(\tau)|\,d\tau\right)^2,$$

where $w_0 = \sqrt{2}/2\cdot\int_0^T \exp(-c(t-\tau)/2)\cdot|e(\tau)|\,d\tau \;/\; (1-\exp(-cT/2))$.

Because $|v(t)-v_2(t)| \longrightarrow 0$ with $t \longrightarrow \infty$ for any solution $v(.)$ of (12) and since (8), (10) and (12) for the function $V_2(t,\mathbf{d}) = W(\mathbf{z}_2(t),\mathbf{d})$ and for any solution $\mathbf{x}(.)$ of (7) we get

$$\limsup_{t \to \infty} \left(V_2(t,\mathbf{x}(t)-\mathbf{z}_2(t)) - v_2(t)\right) \leq 0.$$

Therefore the function E_2 with

$$E_2(t) = \{ x \mid W(t,x-z_2(t)) \le v_2(t)+\varepsilon \}$$

for all $t \in \mathbb{R}$, where ε is a small positive number, defines an attractor localization in $\mathbb{R} \times \mathbb{R}^2$ of system (7). From E_2 and the function W we get for any solution $x(.)$ of (7)

$$\limsup_{t \to \infty} \left(x_1(t)^2 - K_0(t)\right) \le 0,$$

where $K_0(t) = K_0(t;a,b,c,M,\omega) = \left(\sqrt{(a-c^2/4)^2 + 4bv_2(t)} - (a-c^2/4)\right)/b.$
We remark, that $v_2(t) \le 2M^2/c^2$ and $K_0(t;a,b,c,M,\omega) = O(b^{-1/2})$.

The same function $K_0(.)$ we can also get for system (6) by using of the Ljapunov function $V(t,x) = \left(ax_1^2 + cx_1x_2/2 + x_2^2\right)/2 + bx_1^4/4$.

With this Ljapunov function N. P. Bhatia [12] got for $K_0(.)$ a function $K_1(a,b,c,M) = P + bQ = O(b^1)$, where P and Q are positive functions of a, c and M. R. Reissig [16] got by using of geometrical methods a very better bound than K_1: $K_2(a,b,c,M) = 16M^2/(c^2a) = O(b^0)$. Obviously, K_0 is for great parameters b a very better bound than K_1 and K_2.

Construction of the comparison system (3)

Let now be $\mathbf{z}(.) = \bar{\mathbf{x}}(.)$. Then we set $\mathbf{p} = \mathbf{f}(t,\mathbf{z})$ and have $\mathbf{D}(t,\mathbf{z},\mathbf{p}) = 0$ and, therefore, $\gamma_1(t,\mathbf{z},\mathbf{p}) = 0$. Let be

$$g_2(t,\mathbf{z}) = \gamma_2(t,\mathbf{z},\mathbf{f}(t,\mathbf{z})) = -c + \frac{6b\left(z_1[z_2-cz_1+E(t)]\right)_+}{a+3bz_1^2-c^2/4},$$

and

$$g_3(t,\mathbf{z}) = \gamma_3(t,\mathbf{z},\mathbf{f}(t,\mathbf{z})) = $$
$$= b\cdot\left|-1.5c\cdot z_1+z_2-cz_1+E(t)-g_2(t,\mathbf{z})z_1\right|\cdot\left(\frac{2}{a+bz_1^2 - c^2/4}\right)^{3/2}.$$

Let the function E_2 satisfy the conditions of our theorem. Using E_1 and E_2 we determine $\bar{g}_i(t) = \sup_{\mathbf{x} \in E_1(t) \cap E_2(t)} g_i(t,\mathbf{x})$ for $i = 1,2,$ and set $\bar{G}(t,v) = \bar{g}_2(t)\cdot v + \bar{g}_3(t)\cdot v^{3/2}$. The construction of $\bar{G}$ and $\dot{V}(t,\mathbf{d},\mathbf{z}) = W'(t,\mathbf{d},\mathbf{z},\mathbf{f}(t,\mathbf{z}))$ show that the function $F = \bar{G}$ satisfies the condition (α) of our theorem.

Therefore we get a comparison equation (3) in the form

$$\dot{v} = \bar{G}(t,v). \tag{13}$$

A better comparison equation

$$\dot{v} = F(t,v) \tag{14}$$

where $F(t,v) \le \bar{G}(t,v)$ will be obtained, if according to remark 2 the corresponding Supremum is calculated, what actually can be done. In fact:

$$F(t,v) = \sup_{\substack{x \in E_1(t),\ x+d \in E_2(t) \\ V_x(t,d) = v}} \dot{V}(t,x,d) = \sup_{\substack{x \in E_1(t),\ x+d \in E_2(t) \\ V_x(t,d) = v}} -cv + P(t,x,d) \le$$

$$\le -cv + \sup_{x \in E_1(t)} \sup_{\substack{x+d \in E_2(t) \\ V_x(t,d) = v}} P(t,x,d),$$

where $P(t,x,d) = 3b\dot{x}_1 x_1 d_1^2 + b(\dot{x}_1 - c/2x_1)d_1^3 - bc/4d_1^4$ with $\dot{x}_1 = x_2 - cx_1 + E(t)$. Passing to the Supremum with respect to d there is the calculation of Supremum of a cubic polynomial over some interval.

Thus since $F(t,v) \le \overline{G}(t,v)$ the obtained equation (14) possesses better properties of stability, f. e. a larger basin of attraction for the steady position than (13). Since $F(t,v) \le \overline{G}(t,v)$ the system (13) can be also used for the estimate of a basin of attraction for (14) such that remark 1 will work.

Construction of a bassin of attraction for (13)

In order to find a basin of attraction for (13) we notice that by means of the transformation $v = w^2$, $w \ge 0$, the equality (13) can be transformed into the equation

$$\dot{w} = 0.5(\overline{g}_2(t) \cdot w + \overline{g}_3(t) \cdot w^2). \tag{15}$$

Let be $w_1 < - \int_0^T \overline{g}_2(t)\, dt \,/\, \int_0^T \overline{g}_3(t)\, dt$,

$$\kappa = \max_{t \in [0,T]} 0.5 \cdot \int_0^T (\overline{g}_2(\tau) + \overline{g}_3(\tau) \cdot w_1)\, d\tau \quad \text{and}$$

$$w_0 = w_1 \cdot e^{-\kappa}.$$

Then for any solution $w(.)$ of (15) from $w(t_0) \le w_0$ implies the inequality $w(t) \le w_1$ for all $t \ge t_0$ and $w(t) \longrightarrow 0$ for $t \longrightarrow \infty$. Consequently, $\{(t,v) \mid t \in \mathbb{R},\ v \in [0,w_0^2]\}$ is a basin of attraction of (13) and therefore also of (14).

Results

Finally we compare our numerical results with the purely analytic ones of [12]. For this let be $a = c = M = 1$. In [12] (example 3.4) the global asymptotically stability of the periodic solution of (5) is shown for all $b \le 0.028$. This result does not depend on the special choice of the excitation e, since from e only the relation $|e(t)| \le M$ (= 1) can be used.

Let in particular $e(t) = \cos \omega \cdot t$. By means of our proposed method we obtain the following values and upper bounds for the global asymptotically stability of the periodic solution of (5):

$\omega = 0.5$	and	$b \le 922$,
$\omega = 1.0$	and	$b \le 14.2$,
$\omega = 2.0$	and	$b \le 6.52$,
$\omega = 4.0$	and	$b \le 33.2$,

Obviously, these values are much better than the global bound $b \le 0.028$ obtained in [12] and these values depend on ω.

By using of E_2 we determine a upper bound for the Hausdorff dimension of the attractor A of the discrete mapping $\mathbf{T} : \mathbb{R}^2 \longrightarrow \mathbb{R}^2$ with $\mathbf{T}\mathbf{x}_0 = \mathbf{x}(T,\mathbf{x}_0)$, where $\mathbf{x}(.,\mathbf{x}_0)$ is the solution of (7) with $\mathbf{x}(0,\mathbf{x}_0) = \mathbf{x}_0$.

Let be $\mathbf{Q} = \begin{pmatrix} \alpha & -c/2 \\ -c/2 & 1 \end{pmatrix}$, $\mathbf{J}(t,\mathbf{x}) = \frac{\partial}{\partial \mathbf{x}} \mathbf{f}(t,\mathbf{x}) = \begin{pmatrix} -c & 1 \\ -a-3bx_1^2 & 0 \end{pmatrix}$ and $\theta(\mathbf{x}) = c/2 + |a + 3bx_1^2 - \alpha|/[2 \cdot \sqrt{\alpha - c^2/4}\,]$, where $4\alpha > c^2$.

Then for any $t \in \mathbb{R}$ and any $\mathbf{x} \in \mathbb{R}^2$ $\mathbf{QJ}(t,\mathbf{x}) + \mathbf{J}(t,\mathbf{x})^*\mathbf{Q} + 2\theta(\mathbf{x})\mathbf{Q} \ge 0$ holds. Using a theorem of R.A. Smith [17] we get for the Hausdorff dimension $\dim A$ of A the inequality $\dim A < \delta$, when δ satisfies the inequality $0 > \int_0^T \sup_{\mathbf{x} \in E_2(t)} \big((2-\delta)\theta(\mathbf{x}) + \mathrm{tr}\,\mathbf{J}(t,\mathbf{x})\big) dt$, t. e. the inequality

$$\delta > \delta_0 = 2 - 2T \Big/ \int_0^T (1 + \sup_{q \in [0,K_0(t)]} |a+3bq - \alpha|/[c\sqrt{\alpha - c^2/4}\,])\, dt.$$

We get also $\dim A \le \delta_0$. If $\overline{K_0} \ge K_0(t)$ for all $t \in \mathbb{R}$, for example $\overline{K_0} = \big(\sqrt{(a-c^2/4)^2 + 4b\bar{v}} - (a-c^2/4)\big)/b$ with $\bar{v} = 2M^2/c^2$ as an upper bound for the T-periodic solution of (12), and if we choose $\alpha = a + 3/2b\overline{K_0}$, then we get

$$\dim A \le \delta_1 = 2 - 2 \,/\, (\,1 + 3/2b\overline{K_0}/[c\sqrt{\alpha - c^2/4}\,]\,).$$

We remark, that $\dim A = 0$ holds, when $\delta_0 < 1$. In this case the construction of this numbers $\delta_0 \le \delta_1$ over the construction of $\theta(\mathbf{x})$ implies, that there is a global asymptotically stable solution of (7).

For example let be $a = c = M = 1$. Using the above bound $\overline{K_0}$ we get $\delta_1 < 1$ and therefore $\dim A = 0$ and the existence of a global asymptotically stable solution if $b \le 0.312$.

References

[1] Леонов, Г. А.: Метод нелокального сведения в теории абсолютной устойчивости нелинейных систем. I. Автоматика и телемеханика. (1984) 2, 45-53.

[2] Леонов, Г. А.: Метод нелокального сведения в теории абсолютной устойчивости нелинейных систем. II. Автоматика и телемеханика. (1984) 3, 48-56.

[3] Koksch, N.: Anwendung von Ljapunow-Funktionen und Vergleichstechniken auf die Stabilitätsanalyse und Attraktoreingrenzung bei Differentialgleichungssystemen mit kompliziertem Lösungsverhalten. Dissertation A. TU Dresden. Fakultät für Naturwissenschaften und Mathematik. Dresden 1989. 103 S.

[4] Абдуллин, Р. З.; Л. Ю Анапольский; А. А. Воронов; А. С. Землянов; Р. И. Козлов; А. И. Маликов; В. М. Матросов: Метод векторных функций Ляпунова в теории устойчивости. Наука: Москва 1987.

[5] Демидович, Б.П.: Лекции по математической теории устойчивости. Наука: Москва 1967.

[6] Koksch, N.: Untersuchungen zur Stabilität periodischer Systeme mit kompliziertem Lösungsverhalten. Diplomarbeit, TU Dresden Sektion Mathematik, Dresden 1987.

[7] Леонов, Г. А.: Частотный критерий стабилизации нелинейных систем гармоноческим внешним воздействием. Автоматика и телемеханика. (1986) 1, 169-174.

[8] Leonov, G. A.; V. Reitmann: Attraktoreingrenzung für nichtlineare Systeme. Teubner-Texte zur Mathematik Bd. 97, Teubner-Verlag: Leipzig 1987.

[9] Мамедов, Я. Д.; С. Аширов; С. Атдаев: Теоремы о неравенствах. Академия наук Туркменской ССР. Ашхабад 1980.

[10] Воронов, А. А.: Введение в динамику сложных управляемых систем. Наука: Москва 1979.

[11] Rouche, N.; P. Habets; M. Laloy: Stability Theory by Liapunov's Direct Method. Springer: New York - Heidelberg - Berlin 1977.

[12] Bhatia, N. P.: Anwendung der direkten Methode von Ljapunow zum Nachweis der Beschränktheit und der Stabilität der Lösungen einer Klasse nichtlinearer Differentialgleichungen zweiter Ordnung. Dissertation A. TH Dresden, Fakultät für Naturwissenschaften und Mathematik, Dresden 1960.

[13] Неймарк, Ю. И.; П. С. Ланда: Стохастические и хаотические колебания. Наука: Москва 1987.

[14] Reissig, R.; G. Sansone; R. Conti: Qualitative Theorie nichtlinearer Differentialgleichungen. Edizioni Cremonese: Roma 1963.

[15] Schmidt, G.; A. Tondl: Nonlinear Vibrations. Akademie-Verlag: Berlin 1969.

[16] Reissig, R.: Über die asymptotische Stabilität im ganzen bei der Duffingschen Gleichung. Math. Nach. **22**(1960) 129-135.

[17] Smith, R. A.: Some applications of Hausdorff dimension inequalities for ordinary diff. equations. Proc. Roy. Soc. Edingburgh **A 101** (1986) 3-4, 235-259.

ORBITAL STABILITY AND PROBLEMS OF NONLOCAL DYNAMICS

G. A. L e o n o v [1)]

Nowadays it becomes clear, that some classical methods of local analysis which can be traced back to H. Poincaré and A. M. Ljapunov may be used to investigate dynamical systems with complicated behaviour of solutions. One of such methods has been developed primarily by G. Borg [1], P. Hartmann and C. Olech [2,3] to receive sufficient criteria of orbital stability and global stability. Further development of this trend gave the opportunity to formulate the instability criteria [4-6] and to reveal the connection of these theorems [7,8] with the Poincaré theorem [9] and the Andronov-Vitt theorem [9].

In this paper, new results obtained along this way are formulated.

Let us consider the system

$$\frac{dx}{dt} = f(x), \quad x \in R^n, \tag{1}$$

where f is a twice continuously differentiable vector function.

We shall now make a certain remark concerning the terminology. In the process of investigation of strange attractors it turned out to modify to a small extent the notation of orbital stability [9].

Let us introduce the following notations. Suppose $x(.)$ is a trajectory of system (1) and $U_T(x(.),\varepsilon)$ is an ε - neighbourhood of its part, when $t \in [0,T]$, that is

$$U_T(x(.),\varepsilon) = \bigcup_{t \in [0,T]} \{ z \mid |z-x(t)| < \varepsilon \}.$$

In this connection

$$U(x(.),\varepsilon) = \bigcup_{t \in [0,\infty)} \{ z \mid |z-x(t)| < \varepsilon \}.$$

Definition 1. *We say that the trajectory $x(.,x_0)$ is* ***strongly orbitally stable*** *if for arbitrary $\varepsilon > 0$ there exists $\delta > 0$ such that under the condition $|x_0-y_0| < \delta$ the following properties hold:*

1) $x(t,y_0) \in U(x(.,x_0),\varepsilon) \quad \forall\, t \geq 0$,

2) *if* $x(\tau-0,y_0) \in U_T(x(.,x_0),\varepsilon)$ *for certain* $\tau > 0$ *and* $T > 0$ *then*

$$x(\tau,y_0) \in U_t(x(.,x_0),\varepsilon) \quad \forall\, t > T.$$

are valid.

[1)] Leningrad State University, Chair of Theoretical Cybernetics, USSR, 198901 Petrodvorez, Bibliotechnaja pl. 2

Definition 2. *We say that the trajectory* $x(.,x_0)$ *is* ***strongly orbitally unstable*** *if the requirements of Definition 1 are not satisfied.*

It should be noted that the classical definition of orbital stability, which contains only the requirement 1), may "be of no use" for strange attractors in case $x(.,x_0)$ fills the strange attractor densely enough. The requirement 2) gives the opportunity to regard the two trajectories, which diverge from each other with time being, as strongly orbitally unstable. That is why strong orbital unstability may serve as a characteristic of strangeness of attractors.

In this paper and in papers [4,5,6] the orbital unstability is understood in the strong sense.

Suppose at first that $n = 3$ and consider a smooth surface S, placed in the bounded domain $G \subset \mathbb{R}^3$. We shall suppose further that $\mathbf{f}(\mathbf{x}) \neq 0$ in the closure $\bar{G}$ of the domain G.

Denote by $S(\varepsilon)$ an ε-neighbourhood of the surface S and by $\mathbf{n}(\mathbf{x})$ a vector normal to S at the point $\mathbf{x} \in S$. Let $\mathbf{u}(\mathbf{x})$ be a twice continuously differentiable vector function which satisfies the relations $\mathbf{u}(\mathbf{x}) \neq 0$, $\forall\, \mathbf{x} \in \bar{G}$, and $\mathbf{u}^*(\mathbf{x})\mathbf{f}(\mathbf{x}) = \mathbf{u}^*(\mathbf{x})\mathbf{n}(\mathbf{x})$, $\forall\, \mathbf{x} \in S$ ($*$ denotes the transpose).

Let $\mathbf{x}(.)$ and $\mathbf{y}(.)$ be certain trajectories of system (1), which for $t \geq 0$ are contained in $S(\varepsilon)$. Let $\varkappa_1$ and $\varkappa_2$ be certain positive numbers and (t_j) a number sequence which satisfies inequalities $t_{j+1} > t_j$ and $t_{j+1} - t_j < \varkappa_2$. Denote by $\dfrac{\partial \mathbf{f}(\mathbf{u})}{\partial \mathbf{x}}$ the Jacobian matrix of the vector function $\mathbf{f}$ at the point $\mathbf{u}$. Introduce also the notations

$$\varphi(t) = \frac{\mathbf{u}(\mathbf{x}(t))^*}{|\mathbf{u}(\mathbf{x}(t))|^2}\left[\frac{\partial \mathbf{f}(\mathbf{x}(t))}{\partial \mathbf{x}}\mathbf{u}(\mathbf{x}(t)) + \frac{\partial \mathbf{u}(\mathbf{x}(t))}{\partial \mathbf{x}}\mathbf{f}(\mathbf{x}(t))\right],$$

($|.|$ is the Euclidean norm in $\mathbb{R}^n$),

$$\Psi(\mathbf{x}(.),\delta_1,\delta_2) = U(\mathbf{x}(.),\delta_2) \setminus U(\mathbf{x}(.),\delta_1)$$

for any positive numbers δ_1 and δ_2.

Theorem 1. (*on local instability*). *Let for any j the following estimate be true:*

$$\int_{t_j}^{t_{j+1}} \varphi(t)\,dt \geq \varkappa_1, \tag{2}$$

Then there exist continuous functions $\delta_1(\varepsilon)$, $\delta_2(\varepsilon)$ *such that for any* $\mathbf{y}(0) \in \Psi(\mathbf{x}(.),\delta_1(\varepsilon),\delta_2(\varepsilon))$ *it is possible to indicate a number* $\tau = \tau(\varepsilon) > 0$ *which satisfies the relation* $\mathbf{y}(\tau) \notin U(\mathbf{x}(.),\delta_2(\varepsilon))$. *In addition,*

$$\lim_{\varepsilon \to 0} \delta_1(\varepsilon) = 0, \qquad \lim_{\varepsilon \to 0} \delta_2(\varepsilon) > 0. \tag{3}$$

Theorem 2. *(on local stability). Let for any j the following estimate be true:*

$$\int_{t_j}^{t_{j+1}} \varphi(t)dt \le -\varkappa_1, \tag{4}$$

Then there exist continuous functions $\delta_1(\varepsilon)$ and $\delta_2(\varepsilon)$ such that for any $y(0) \in U(x(.),\delta_2(\varepsilon))$ it is possible to indicate a number $\tau > 0$ with $y(t) \in U(x(.),\delta_1(\varepsilon))$, $\forall\, t \ge \tau$. In addition,

$$\lim_{\varepsilon \to 0} \delta_1(\varepsilon) = 0, \qquad \lim_{\varepsilon \to 0} \delta_2(\varepsilon) > 0. \tag{5}$$

Suppose now that $|u(x(t))| = 1$, $\forall\, t \ge 0$. In this case $\varphi(t) = u(x(t))^* \dfrac{\partial f(x(t))}{\partial x} u(x(t))$ and it is possible to formulate the following two assertions.

Theorem 3. *Let the inequality (2) be true and $y(t_j) \in \Psi(x(.),\delta_1(\varepsilon),\delta_2(\varepsilon))$ for $j < N$. Then*

$$\frac{|y(t_N) - x(t_N)|^2}{|y(0) - x(0)|^2} = \exp 2 \int_0^{t_N} \varphi(t)dt \,(1 + O(\delta_2(\varepsilon),\varepsilon)).$$

Here $\delta_1(\varepsilon)$, $\delta_2(\varepsilon)$ are defined in the text of theorem 1.

Suppose now that there exists a limit

$$\lim_{t \to +\infty} \frac{1}{t} \int_0^{t} \varphi(t)dt = \lambda.$$

Theorem 4. *Suppose the inequality (2) is true. Suppose also that $y(t_j) \notin \Psi(x(.),\delta_1(\varepsilon),\delta_2(\varepsilon))$ for $j < N(\varepsilon)$. Then $N(\varepsilon) \to 0$ as $\varepsilon \to 0$ and*

$$\lim_{\varepsilon \to 0} \frac{1}{t_{N(\varepsilon)}} \ln \frac{|y(t_{N(\varepsilon)}) - x(t_{N(\varepsilon)})|}{|y(0) - x(0)|} = \lambda.$$

When $\varepsilon = 0$ theorems 1 and 2, respectively, turn into the theorems about orbital instability and about stability of trajectories on the integral surface S. When S is a plane, $x(.)$ is a T-periodic trajectory, and $t_{j+1} - t_j = T$, then these criteria coincide with a well-known Poincaré test [9]. Indeed, in this case with the help of a unitary mapping $x = Q\tilde{x}$ ($Q^* = Q^{-1}$) vector functions f and u, respectively, can be brought to the form

$$\tilde{f}(\tilde{x}) = Q^{-1}f(Q\tilde{x}) = \begin{Vmatrix} \tilde{f}_1(\tilde{x}) \\ \tilde{f}_2(\tilde{x}) \\ 0 \end{Vmatrix}, \qquad \tilde{u}(\tilde{x}) = Q^{-1}u(Q\tilde{x}) = \begin{Vmatrix} \tilde{u}_1(\tilde{x}) \\ \tilde{u}_2(\tilde{x}) \\ 0 \end{Vmatrix}.$$

If $\tilde{u}_1 = \tilde{f}_2$, $\tilde{u}_2 = -\tilde{f}_1$ the left parts of inequalities (2) and (4) take the form

$$\int_0^T \left[\frac{\partial f_1(\tilde{x}(t))}{\partial \tilde{x}_1} + \frac{\partial f_2(\tilde{x}(t))}{\partial \tilde{x}_2} \right] dt.$$

This fact means that inequalities (2) and (4) coincide with the Poincaré test.

Let us discuss[2)] the way of application of these results to the Lorenz equation [10]

$$\dot{x}_1 = -\sigma x_1 + \sigma y_1, \; \dot{y}_1 = rx_1 - y_1 - x_1 z_1, \; \dot{z}_1 = -bz_1 + x_1 y_1 \tag{6}$$

It is well-known [10] that if $\sigma = 10$, $r = 28$, $b = 8/3$ then in system (6) by means of numerical experiments a strange attractor Φ, which is near to two-dimensional surface, is observed. Let us choose a certain point with coordinates x_{10}, y_{10}, z_{10} on the attractor Φ and emit a trajectory $x(.)$ from it. Let the set L consist of points with coordinates x_{10}, y_{10}, z_{10} and $-x_{10}$, $-y_{10}$, z_{10}. Let us determine also a sequence $t_j \to +\infty$ as $j \to \infty$, such that the distance between $x(t)$ and the set L does not exceed 10^{-2}. In the process of numerical construction of $x(t)$ and $\varphi(t)$ it comes out that

$$\int_{t_j}^{t_{j+1}} \varphi(t)dt > 0.$$

Therefore, by means of theorem 1 we may conclude that trajectories, which are disposed in the attractor of system (6), are locally instable.

The value λ from the text of theorem 4 is an analogue of Ljapunov characteristic exponent [11,12]. Therefore, when $\operatorname{div} f(x) = -\alpha$, $\forall x \in \mathbb{R}^3$, where α is a positive number it is possible to introduce the notion of Ljapunov dimension by the Kaplan-Yorke formula [12,13]

$$d_L = 2 + \frac{\lambda}{\lambda + \alpha}.$$

For $\sigma = 16$, $b = 4$, $r = 40$, with the help of theorem 4, the numerical calculations give $d_L = 2.056$. This result is near $d_L = 2.06$ which is obtained in [15].

Now, we shall formulate two lemmas. From them theorems 1-3 follow. To this end, we shall consider system (1), where $x \in \mathbb{R}^n$, provided that $f(x) \neq 0$ in $\bar{G} \subset \mathbb{R}^n$.

---

2) Numerical computation of $x(t)$ and $\varphi(t)$ for system (6) has been done by V. L. Chistjakov.

Let us introduce a symmetric nonsingular matrix $H(x) = \| h_1(x), \ldots, h_n(x) \|$, where $h_i(x)$ are twice continuously differentiable vector functions. Let us introduce also a twice continuously differentiable vector function $q(x)$ which satisfies the relation $f(x)^* q(x) \neq 0$, $\forall\, x \in \bar{G}$. We shall suppose that the quadratic form $z^* H(x(t)) z$ is non-degenerate on the set $\{ z \mid z^* q(x(t)) = 0 \}$. Suppose also that for a certain symmetric matrix H_0 the inequality $z^* H(x(t)) z \geq z^* H_0 z$ is true on the set mentioned.

Introduce the notation

$$\left(\frac{\partial H(x)}{\partial x},\ f(x) \right) = \left\| \frac{\partial h_1(x)}{\partial x} f(x),\ \ldots,\ \frac{\partial h_n(x)}{\partial x} f(x) \right\|,$$

where $\frac{\partial h_i(x)}{\partial x}$ is the Jacobian matrix of the vector function $h_i(x)$ at the point x.

For a fixed number K and sufficiently small numbers δ consider the set

$$\Omega(\delta) = \bigcup_{t \geq 0} \left\{ y \mid (y-x(t))^* H(x(t))(y-x(t)) = \delta,\ |y-x(t)|^2 \leq K|\delta|,\ (y-x(t))^* q(x(t)) = 0 \right\}.$$

Lemma 1. *For any point $y_0 \in \Omega(\delta)$ there exists a number $t \geq 0$, such, that*

$$l(y_0)^* f(y_0) = (y_0-x(t))^* \Big\{ \frac{1}{2} \Big(\frac{\partial H(x(t))}{\partial x},\ f(x(t)) \Big) + H(x(t)) \frac{\partial f(x(t))}{\partial x} - H(x(t)) f(x(t)) \left[f(x(t))^* q(x(t)) \right]^{-1} \cdot \Big[f(x(t))^* \Big(\frac{\partial q(x(t))}{\partial x} \Big)^* + q(x(t))^* \Big(\frac{\partial f(x(t))}{\partial x} \Big)^* \Big] \Big\} (y_0-x(t)) + O(|\delta|^{3/2}).$$

Here, the vector $l(y_0)$ is the normal to $\Omega(\delta)$ in the point y_0 and it is such that

$$\Big(I - \frac{q(x(t))q(x(t))^*}{|q(x(t))|^2} \Big) l(y_0) = \Big(I - \frac{q(x(t))q(x(t))^*}{|q(x(t))|^2} \Big) H(x(t))(y_0-x(t)).$$

Lemma 2. *For any point $y_0 \in \Omega(\delta)$ for small τ*

$$y(\tau, y_0) \in \Omega(\delta + \tau \cdot l(y_0)^* f(y_0) + O(\tau^2)).$$

Here, $y(\cdot, y_0)$ is a solution of system (1) with the initial condition $y(0, y_0) = y_0$.

The proofs of lemma 1 and lemma 2 can be deduced from the arguments of paper [7].

Notice now, that in order to prove theorem 1 it is sufficient to put $H(x) = -I \cdot |u(x)|^2$ and in order to prove theorem 2 we can put $H(x) = I \cdot |u(x)|^2$. In this connection $q(x) = f(x)$.

Consider then the set $\Omega(\delta) \cap S(\varepsilon)$ from lemma 1. We receive that for $y_0 \in \Omega(\delta) \cap S(\varepsilon)$ there exists a number t, such that $l(y_0)^* f(y_0) = 2\delta\varphi(t) + O(|\delta|^{3/2}, \varepsilon)$.

From this and from lemma 2 it follows that for small δ and ε

$$y(t,y_0) \in \Omega\Big(\exp 2\int_0^t \varphi(t)dt \Big[\delta + O(|\delta|^{3/2},\varepsilon)\int_0^t \exp \int_s^0 \varphi(\tau)\, d\tau ds\Big]\Big). \qquad (7)$$

It follows from (7) that the inclusion

$$y(t,y_0) \in \Omega\Big(\delta \cdot \exp 2\int_0^t \varphi(t)dt + O(|\delta|^{3/2},\varepsilon)\Big) \qquad (8)$$

is also fulfilled. With the help of (8) and the standard Ljapunov techniques [3,9] one can get now the statements of theorem 1 and 2. theorem 3 arises from (9) and theorem 4 is a corollary of theorem 3.

Let us pass now to the theorem on global asymptotic stability of system (1) when the system is dissipative and has several equilibria. We shall assume for definiteness that system (1) has two asymptotically stable equilibria and one saddle, at which the Jacobian matrix $\frac{\partial f(x)}{\partial x}$ has one positive and $(n-1)$ negative eigenvalues.

Let $\lambda_1(x) \geq \lambda_2(x) \geq \ldots \geq \lambda_n(x)$ be eigenvalues of the matrix $\Big[\frac{\partial f(x)}{\partial x} + \frac{\partial f(x)}{\partial x}^* \Big]/2$ and D be the dissipation domain of system (1).

<u>**Theorem 5.**</u> *Suppose there exists $\varkappa > 0$, such that the inequality*

$$\int_{t_j}^{t_{j+1}} \big[\lambda_1(x(t)) + \lambda_2(x(t)) \big] dt \leq \varkappa, \ \forall j \qquad (10)$$

is valid on every solution $x(t) \in D$. Then system (1) is completely stable, i.e. its any solution tends to an equilibrium as $t \to +\infty$.

This theorem is a certain generalization and modification of the Hartman-Olech theorem [2,3] and Smith theorem [15]. Its proof is based on the ideas of the papers [7,16].

Let us denote by G_1 and G_2 domains of attraction for the stable equilibria. Let $x_0 \in \partial G_1$ where ∂G_1 is the boundary of G_1.

Notice that when the ω-limit set of $x(.)$ does not contain any equilibria it follows from (10) that $x(.)$ is uniformly asymptotically orbitally stable [16]. Therefore for the trajectory $x(.,x_0)$ the saddle equilibrium x_3 will be an ω-limit point.

Suppose there exists another ω-limit point $x_4 \neq x_3$ and for the sequence $t_k \to +\infty$ and $t_l \to +\infty$ the relations

$$\lim_{k \to \infty} x(t_k, x_0) = x_3, \qquad \lim_{l \to \infty} x(t_l, x_0) = x_4$$

are carried out. Let us consider a convergent sequence of curves $x(t,x_0)$, $t \in [t_l, t_k]$. It is clear that it converges to $x(t,x_4)$ and, consequently, $x(t,x_4) \to x_3$ as $t \to \infty$.

In virtue of the invariance of ω-limit sets the trajectory $x(.,x_4)$ is entirely arranged in the ω-limit set of $x(.,x_0)$. Consequently $x(t,x_4) \in D$, $\forall\, t \in \mathbb{R}$. If the α-limit set of $x(.,x_4)$ is separated from x_3 then in virtue of (10) it consists of uniformly asymptotically orbitally stable trajectories [16]. But this fact contradicts the relation $\lim_{l \to \infty} x(t_l,x_4) = x_3$ as $t \to -\infty$.

Thus $x(.,x_4)$ is a separatrix loop of system (1). Let us span now on this loop a uniformly smooth two-dimensional surface L in such a way that $x(.,x_4)$ is placed on the boundary of L. Consider $x(k,L)$. From this sequence we can single out a subsequence which converges to a two dimensional surface M. It is clear that $x(t,M) = M$. From this fact and from (10) it follows according to Corollary 2.1 [15] that the separatrix loop is impossible.

Thus $x(t,x_0) \to x_3$ as $t \to +\infty$. Using this and the fact that the stable manifold of saddle x_3 is homeomorphic to a hyperplane it follows that any solution of (1) converges to an equilibrium.

Applying now theorem 5 to system (6) we shall receive the following condition of complete stability in the case $b \geq 1$.

Theorem 6. *Suppose the following conditions are fulfilled:*

(*i*) *in case* $b \geq 2$

$$(\sigma+b)(b+1) > \frac{r\sigma}{2}\left[1 + \sqrt{\frac{(\sigma+b)(b-2)^2}{(\sigma+1)4(b-1)} + 1}\right],$$

$$\sqrt{(\sigma+1)\left[\frac{b^2(b+\sigma)}{4(b-1)} - b + 1\right]} \geq b - 1,$$

(*ii*) *in case* $b \in [1,2]$

$$(\sigma+b)(b+1) > r\sigma, \quad \sigma+2 \geq b.$$

Then any solution of system (6) *tends to an equilibrium as* $t \to +\infty$.

If $b = 8/3$, $\sigma = 10$ then the hypotheses of theorem 6 are carried out provided that $r \leq 4.5$.

Mathematical methods which are developed to obtain orbital stability and instability criteria can be applied with the aim to evaluate both Ljapunov and Hausdorff dimension of attractors. In this way, we shall get an estimate for Hausdorff dimension dim A of an attractor A (i.e. of an attracting invariant compactum) of the system (1) which is more precise in a number of cases than the estimates from the articles [15,17]. Suppose also that $\mathbf{f}(\mathbf{x}) \neq \mathbf{0}\ \forall\ \mathbf{x} \in A$.

Let $J(\mathbf{p}) = \frac{\partial \mathbf{f}}{\partial \mathbf{x}}(\mathbf{p}) + \left(\frac{\partial \mathbf{f}}{\partial \mathbf{x}}(\mathbf{p})\right)^*$ and $\lambda_j(\mathbf{x})$ $(j = 1,\dots,n-1)$ be the eigenvalues of the quadratic form $\mathbf{y}^* J(\mathbf{x})\mathbf{y}$ matrix on the set $\{\ \mathbf{y}\ |\ \mathbf{y}^*\mathbf{f}(\mathbf{x}) = 0\ \}$. Let $\lambda_1(\mathbf{x}) \geq \lambda_2(\mathbf{x}) \geq \dots \geq \lambda_{n-1}(\mathbf{x})$.

Theorem 7. *Suppose that there exist an integer l and a number $s \in [0,1]$ such that the inequality*

$$\lambda_1(\mathbf{x}) + \dots + \lambda_{l-1}(\mathbf{x}) + s\lambda_l(\mathbf{x}) < 0,\ \forall\ \mathbf{x} \in A \tag{11}$$

holds. Then dim $A < l + s$.

In order to prove the theorem we shall use the following notations.

Suppose that for the numbers t and τ $|t - \tau|$ is small. Then, for a solution $\mathbf{x}(.)$ of system (1) we have

$$\mathbf{x}(\tau) \approx \mathbf{x}(t) + \mathbf{f}(\mathbf{x}(t))(t-\tau).$$

Let us define a mapping $\mathbf{Q} : \Omega_1 \longrightarrow \Omega_2$ with

$$\mathbf{Q}(\mathbf{v}) = \mathbf{v} + \alpha\left[\ \mathbf{f}(\mathbf{x}(t)) + \frac{\partial \mathbf{f}}{\partial \mathbf{x}}(\mathbf{x}(t))(\mathbf{v}-\mathbf{x}(t))\ \right],$$

where Ω_1 is the hyperplane $\Omega_1 = \{\ \mathbf{v}\ |\ (\mathbf{v}-\mathbf{x}(t))^*\mathbf{f}(\mathbf{x}(t)) = 0\ \}$ and Ω_2 is the hyperplane

$$\Omega_2 = \{\ \mathbf{u}\ |\ [\ \mathbf{u} - \mathbf{x}(t) + (\tau-t)\mathbf{f}\ ]^*[\ \mathbf{f} + (\tau-t)\mathbf{K}\mathbf{f}\ ] = 0\ \}.$$

Here

$$\mathbf{f} = \mathbf{f}(\mathbf{x}(t)),\quad \mathbf{K} = \frac{\partial \mathbf{f}}{\partial \mathbf{x}}(\mathbf{x}(t)) \quad \text{and}$$

$$\alpha = (\tau-t)\left[\ |\mathbf{f}|^2 - \mathbf{f}^*\mathbf{K}^*(\mathbf{v}-\mathbf{x}(t)) + (\tau-t)\mathbf{f}^*\mathbf{K}\mathbf{f}\ \right]\cdot$$
$$\cdot\left[\ |\mathbf{f}|^2 + \mathbf{f}^*\mathbf{K}(\mathbf{v}-\mathbf{x}(t)) + (\tau-t)\mathbf{f}^*\mathbf{K}\mathbf{f} + (\tau-t)\mathbf{f}^*\mathbf{K}\mathbf{K}(\mathbf{v}-\mathbf{x}(t))\ \right]^{-1}$$

It is clear that when $(\tau-t)$ and $(\mathbf{v}-\mathbf{x}(t))$ are small, the mapping $\mathbf{Q}$ is the linear approximation of the shift mapping of Ω_2 into Ω_2 along the trajectories of system (1). It is clear also that in this case

$$\mathbf{Q}(\mathbf{v}) - [\mathbf{x}(t) + (\tau-t)\mathbf{f}(\mathbf{x}(t))] \approx \mathbf{M}(\mathbf{v}-\mathbf{x}(t)).$$

Here $\mathbf{M} = \mathbf{I} + (\tau-t)\left[\frac{\partial \mathbf{f}}{\partial \mathbf{x}}(\mathbf{x}(t)) - \frac{\mathbf{f}(\mathbf{x}(t))\mathbf{f}(\mathbf{x}(t))^*}{|\mathbf{f}(\mathbf{x}(t))|^2}\left(\frac{\partial \mathbf{f}}{\partial \mathbf{x}}(\mathbf{x}(t)) + \left(\frac{\partial \mathbf{f}}{\partial \mathbf{x}}(\mathbf{x}(t))\right)^*\right)\right]$.

Let us consider now a certain positive definite matrix $\mathbf{H}$, a number δ and the ellipsoid $E = \{\ \mathbf{z}\ |\ \mathbf{z}^*\mathbf{H}\mathbf{z} = \delta,\ \mathbf{z}^*\mathbf{f}(\mathbf{x}(t)) = 0\ \}$ located in the hyperplane $\{\ \mathbf{z}\ |\ \mathbf{z}^*\mathbf{f}(\mathbf{x}(t)) = 0\ \}$. The ellipsoid $\mathbf{M}E$ is the image of the ellipsoid E under the linear mapping $y = Mz$ of the hyperplane $\{ z \mid z^* f(x(t)) = 0 \}$ into the hyperplane $\{ y \mid f(x(t))^* M^{-1} y = 0 \}$.

Let us denote by $a_1(E),\dots,a_l(E)$ and by $a_1(ME),\dots,a_l(ME)$ the lengths of the semi-axes of the ellipsoids E and ME respectively. Suppose that both sequences are decreasing. Let

$$\omega(E) = a_1(E)\cdot\ldots\cdot a_{l-1}(E)a_l(E)^s$$

and

$$\omega(ME) = a_1(ME)\cdot\ldots\cdot a_{l-1}(ME)a_l(ME)^s.$$

Lemma 3. *If the inequality* (11) *holds then* $\omega(ME) < \omega(E)$.

Proof. Let us consider the matrix

$$L = \frac{\partial f}{\partial x}(x(t)) - \frac{f(x(t))f(x(t))^*}{|f(x(t))|^2}\left(\frac{\partial f}{\partial x}(x(t)) + \left(\frac{\partial f}{\partial x}(x(t))\right)^*\right) + d_1(f(x(t)))^*,$$

where the vector d_1 is chosen in such a way that the first l greatest eigenvalues of the matrix $L + L^*$ coincide with $\lambda_1(x(t)),\dots,\lambda_l(x(t))$. It is clear that the operator $y = (I + (\tau-t)L)z$ is a certain extension on the whole space $\mathbb{R}^n$ of the operator $y = Mz$ which is defined on the hyperplane $\{ z \mid z^* f(x(t)) = 0 \}$.

Let us consider also a certain ellipsoid

$$\tilde{E} = \{ z \mid z^*(M + d_2 x^*(t))z = \delta \}$$

in the space $\mathbb{R}^n$, where d_2 is chosen in such a way that the greatest l semi-axes of $\tilde{E}$ are located in $\{ z \mid z^* f(x(t)) = 0 \}$. So, $\omega(E) = \omega(\tilde{E})$.

Let us denote by $\mu_1 \geq \mu_2 \geq \ldots \geq \mu_n$ the eigenvalues of the matrix $(I+(\tau-t)L)^*(I+(\tau-t)L)$.

It is well known [15,17] that from $\mu_1\cdot\ldots\cdot\mu_{l-1}\mu_l^s < 1$ it follows that

$$\omega((I+(\tau-t)L)\tilde{E}) < \omega(\tilde{E}). \tag{12}$$

But as $|\tau-t|$ is small we have

$$\mu_1\cdot\ldots\cdot\mu_{l-1}\mu_l^s \approx 1 + (\lambda_1(x) + \ldots + \lambda_{l-1}(x) + s\lambda_l(x))(\tau-t).$$

Using this and (11) we get (12). But as $\omega(E) = \omega(\tilde{E})$ and

$$\omega((I+(\tau-t)L)\tilde{E}) \geq \omega(ME)$$

we have finally $\omega(ME) < \omega(E)$.■

Consider now the trajectory $x(.)$ of (1) with $x(t) \in A$ for all t and the smooth $(n-1)$-dimensional surfaces $P(x(t))$ which pass through the points $x(t)$ and have normal vectors that coincide with $f(x)$, $x \in P(x(t))$.

We denote by $\mu(A \cap P(x(t)),d)$ the elliptic Hausdorff d-measure [17] of the set $A \cap P(x(t))$.

Lemma 4. *If $d > l + s - 1$ and the inequality (11) holds then $\mu(A \cap P(x(t)),d)$ strictly monotonically decreases along the trajectory $x(.)$.*

Lemma 4 is a direct corollary of lemma 3.

In order to prove theorem 7 it is sufficient now to notice that as A is a compactum and $x(t) \in K$, $\forall\, t \in \mathbb{R}$, then there exists an ω-limit point x_0 of the trajectory $x(.)$. Let us choose numbers $t_1 < t_2$ in such a way that $x(t_1)$ and $x(t_2)$ are disposed in the sufficiently small neighborhood of x_0 and $P(x(t_1)) = P(x(t_2))$. It follows from lemma 4 that for all $d > l + s - 1$ the inequality $\mu(A \cap P(x(t)),d) = 0$ holds. But then $\dim A < l + s$.

Notice that different generalizations of theorem 7 are possible by using the surfaces $P(x(t))$ which have normal vectors $q(x)$ that do not coincide with $f(x)$ in any points x of $P(x(t)$ but only satisfy the inequality $f(x(t))^* q(x(t)) \neq 0$. The analogies of lemma 3 in this case may be obtained with the help of the methods developed in [6].

Notice also that the inequality (12) may be made more precisely in the following way [17] $\omega((I+(\tau-t)L)\tilde{E}) \leq \mu_1 \cdot \ldots \cdot \mu_{l-1}\mu_l^s \omega(\tilde{E})$. Then

$$\dot{\omega}(\tilde{E}) = \lim_{\tau \to t} \frac{\omega((I+(\tau-t)L)\tilde{E}) - \omega(\tilde{E})}{\tau - t} \leq$$

$$(\lambda_1(x(t)) + \ldots + \lambda_{l-1}(x(t)) + s\lambda_l(x(t)))\omega(\tilde{E}).$$

That is why theorem 7 holds if the inequality (11) is replaced by the inequality

$$\int_{t_1}^{t_2} \lambda_1(x(t)) + \ldots + \lambda_{l-1}(x(t)) + s\lambda_l(x(t))\, dt < 0. \qquad (13)$$

Here $x(.)$ is an arbitrary trajectory from A and $t_1 < t_2$ are arbitrary numbers such that $P(x(t_1)) = P(x(t_2))$.

Now, let us consider the Lorenz system (6), where $\sigma = 16$, $b = 4$ and $r = 40$.

It is well known from papers [18-20] that in this case the Lorenz system has a strange attractor A. This attractor has been modeled by means of computer[3]. Applying these computer results and the integral inequality (13) the estimate of dim A near to 2 is obtained. This result is near to that of the papers [19,20].

---

[3] It has been done by D. V. Ponomarenko and V. L. Chistjakov

References

[1] Borg, G.: A condition for the existence of orbitally stable solutions of dynamical systems. Kungl. Tekn. Högsk. Handl. Stockholm. **158**(1960), 12 pp.

[2] Hartman, P., and C. Olech, C.: On global asymptotic stability of solutions of differential equations. Trans. Amer. Math. Soc. **104**(1962), p. 154-176.

[3] Hartman, P.: Ordinary differential equations. Moscow, 1970.

[4] Leonov, G.A. (Леонов, Г.А.): Orbital stability of nonlinear systems trajectories. Вестник ЛГУ (1987)1, p. 26-29. [Russian]

[5] Leonov, G.A. (Леонов, Г.А.): On orbital stability of autonomous systems trajectories. Дифф. урав. (1988)4, p. 694-695. [Russian]

[6] Leonov, G.A., and V. Reitmann: Attraktoreingrenzung für nichtlineare Systeme. Teubner-Texte zur Mathematik, Bd. 97, Leipzig 1987.

[7] Leonov, G.A. (Леонов, Г.А.): On a higher-dimensional analogue of the Poincaré test for orbital stability. Дифф. урав.(1988)9, p. 1637-1639.

[8] Boichenko, V.A. (Бойченко, В.А.), and G.A. Leonov (Леонов, Г.А.): On orbital Ljapunov exponents for autonomous systems. Вестник ЛГУ (1988)3, p.7-10. [Russian]

[9] Demidovich, B.P. (Демидович, Б.П.): Lectures on mathematical stability theory. Moscow 1967.

[10] Strange attractors. Collection of papers. Moscow 1981.

[11] Bybov, B.F. (Бывов, Б.Ф.), R.E. Vinograd (Виноград, Р.Е), D.M. Grobman (Гробман, Д.М.), and V.V. Nemytskij (Немыцкий, В.В.): Ljapunov indexes theories. Moscow 1966. [Russian]

[12] Neimark, Ju.I. (Неймарк, Ю.И.), and P.S. Landa (Ланда, П.С.): Stochastic and chaotic oscillations. Moscow 1987.

[13] Kaplan, J.L., and J.A. Yorke: Chaotic behavior of higherdimensional difference equations. Lect. Notes in Math. N. 730, 1979, p. 204-227.

[14] Mori, H., and H. Fujisaka: Statistical dynamics of turbulence. Lect. Notes in Phys. N. 132, 1980, p. 181-197.

[15] Smith, R.A.: Some applications of Hausdorff dimension inequalities for ordinary differential equations. Proc. Roy. Soc. Edinburgh, **A 104** (1986) 3-4, p. 235-259.

[16] Leonov, G.A. (Леонов, Г.А.): On orbital stability of dynamical systems trajectories. Прикл. мат. и мех. (to appear). [Russian]

[17] Douady, A. and J. Oesterlé: Dimension de Hausdorff des attracteurs. C.R.Acad.Sc. Paris. Sér. A **290** (1980), p. 1135-1138.

[18] More, H. and H. Fujisaka: Statistical dynamics of turbulence. Lect. Notes in Phys. v. 132, Berlin, Springer-Verlag 1980, p. 181-197.

[19] Mc Guinness, M. J.: Computation of the limit capacity of the Lorenz attractor. Physica D. **16**(1985)2, p. 265-275.

[20] Mc Guinness, M. J.: The fractal dimension of the Lorenz attractor. Phys. Lett. **99A**(1983)1, p. 5-9.

ABOUT THE STRUCTURE OF CONSERVED QUANTITIES OF SPECIAL MANY-BODY SYSTEMS

P. MOEBIUS

Department of Physics, Technical University of Dresden, GDR

1. Introduction

In the last decade an enormous progress was achieved in the investigation of nonlinear dynamical systems, being represented by a set of nonlinear differential equations. It has become customary to split the problems into "integrable" and "nonintegrable" ones with the idea behind it, that in the first case they are well defined and reliable methods exist, to treat them systematically while in the second case no reliable analytical or numerical procedures are at present available to get conclusive results. An essential ingredient of integrable systems of Hamiltonian classical mechanics is the existence of a sufficient number of conserved quantities [1]. But it is allways tedious to find them and no systematic way exists so far to calculate them straightforwardly. Some "inverse method" for special translationally invariant many-particle systems will be pesented with the content to postulate a definite structure of conserved quantities and to ask, which class of problems can be assigned to it. This task leads to general conditions for the admitted potentials in the form of differential equations or addition theorems. It will be demonstrated that the Hamiltonian formalism is not very appropiate for looking at the conserved quantities or invariant sets belonging to the problem. It was developed to treat the time-evolution, e.g. the orbits and phase portraits. But for many purposes this is not important, e.g. statistical physics, but the invariant sets, flows or even manifolds are more informative. So a way of describing integrable systems will be presented, indicating that other physical quantities might be preferable for their characterization, e.g. the momenta and square roots of the two body potentials.

2. Integrable systems

Let us shortly repeat the essential facts about integrable Hamiltonian systems. They can be described by a Hamiltonian H depending on 2f canonically conjugate variables (p,q), where f is a finite number of degrees of freedom, possessing f conserved quantities $J_i(p,q)$ with

$$\dot{J}_i(p,q) = \{ H, J_i \} = 0, \qquad i = 1 \,..\, f, \tag{1a}$$

where these quantities have to be in involution, which in the formalism of Poisson brackets means

$$\{ J_i, J_j \} = 0, \qquad i,j = 1 \,..\, f, \tag{1b}$$

and where the dJ_i are linearily independent [1]. In the following we shall concentrate us on possible relations between the form of the interaction potential and the structure of the conserved quantities in a way that the last ones should be connected with certain (algebraic) invariants of a mathematical structure characterizing the system.

The following considerations are restricted to the case of a system with f one dimensional particles on a ring, i.e. with periodic boundary conditions of the type $q_{f+1} = q_1$, having a translationally invariant interaction potential and a Hamiltonian of the type

$$H = \sum_k \frac{p_k^2}{2m} + \sum_{k,l} V(q_k - q_l), \qquad k,l = 1 \,..\, f. \tag{3}$$

The task is now to find a systematic way

i.) informing us for which class of potentials $V(q_k, q_l)$ is the system (3) integrable,

ii) providing us in a straightforward way with the necessary conserved quantities,

iii) relating a given structure of conserved quantities with the corresponding class of potentials.

Within the Hamiltonian formalism it seems to be rather tedious to find a suitable way for solving this task. Therefore a new method, basing on an "inverse procedure", shall be presented. To make it clear, it seems very appropiate to start with the case of 3 degrees of freedom.

3. Examples with 3 degrees of freedom

3.1. 3-dimensional harmonic oscillator

A well know example of f = 3 is the isotropic harmonic oscillator with the Hamiltonian

$$H = \frac{1}{2m}(p_1^2 + p_2^2 + p_3^2) + \frac{1}{2} c(q_1^2 + q_2^2 + q_3^2). \tag{4}$$

It is advantageous to introduce the quantities

$$a_k = \frac{p_k}{\sqrt{2m}} + iw_k, \quad a_k^* = \frac{p_k}{\sqrt{2m}} - iw_k, \quad \text{with } w_k = \sqrt{\frac{c}{2}}\, q_k \tag{5a}$$

and the matrix

$$M = (a_k^* a_l), \qquad k,l = 1, 2, 3, \tag{5b}$$

which can be invariantly decomposed into 3 parts accordingly $3 \times 3 = 1 \oplus 3 \oplus 5$;

$$M = \frac{1}{3} I\, \mathrm{tr} M + M^{(a)} + M^{(s)}, \tag{6a}$$

where I is the 3 x 3 unit matrix, trM = H and $M^{(a)}$ is totally skew symmetric

$$M_{kl}^{(a)} = -M_{kl}^{(a)}, \quad M_{kl}^{(a)} = \frac{1}{2}(a_k^* a_l - a_l^* a_k) = \frac{i}{2}(p_k w_l - p_l w_k) \tag{6b}$$

being proportional to the 3 components of the angular momentum, and $M^{(s)}$ is the totally symmetric traceless matrix:

$$M_{kl}^{(s)} = M_{lk}^{(s)} \text{ with } \mathrm{tr} M^{(s)} = 0, \quad M_{kl}^{(s)} = \frac{1}{2}(a_k^* a_l + a_l^* a_k) - \frac{1}{3}\delta_{kl}\, \mathrm{tr} M, \quad k,l = 1 \,..\, f. \tag{6c}$$

which is proportional to the totally symmetric orbital angular momentum tensor. Now all these quantities are time-independent:

$$\dot{H} = \mathrm{tr}\dot{M} = 0, \quad \dot{M}_{kl}^{(a)} = 0, \quad \dot{M}_{kl}^{(s)} = 0, \tag{7}$$

but there exists the following relations between them

$$H^2 = \frac{3}{2}\left(\sum_{kl} M_{kl}^{(a)2} + \mathrm{tr}(M^{(s)} M^{(s)})\right). \tag{8}$$

So we have in total 9 time-independent quantities

$$(a_k^{\times} a_l)^{\cdot} = 0. \tag{9}$$

The question is now how many of them are independent and in involution. The symmetry group of a 3-dimensional harmonic oscillator is SU(3), having 8 generators which are connected with the quantities $M^{(a)}$ (3) and $^{(s)}M$ (5), possessing a maximal set of 5 commuting operators [2]. But due to the relation (8) we have here only 4 independent conserved quantities which are in involution. One can choose e.g. the set

$$H, \quad M_{12}^{(a)}, \quad \sum_{i,j} M_{ij}^{(a)2}, \text{ one component of } M_{kl}^{(s)}, \tag{10}$$

The general statement is now that the conserved quantities are bilinear forms in the canonically conjugate variables (p_k, q_k)

$$J_n = \sum_{v,i,j} a_{ij}^{n,v} \, p_i^{2-v} \, q_j^v, \quad v = 0,1,2; \; i,j = 1,2,3; \; n = 1,2,3,4, \tag{11}$$

which can also be formulated in the "appropriate variables" p_k, w_k due to the relation (5a), being very essential in the following.

3.2. Harmonic and anharmonic coupling of 3 particles

We will now consider the case of 3 one-dimensional particles on a ring, i.e. with periodic boundary conditions, with the following type of coupling

$$H = \sum_i^3 \frac{1}{2m} p_i^2 + \sum_{i,j=1}^3 \frac{c}{2}(q_i-q_j)^2 + k(q_i-q_j)^4. \tag{12}$$

For the further treatment it is advantageous to introduce new coordinates by the orthogonal transformation [3]

$$\begin{aligned} q_1' &= \frac{3-\sqrt{3}}{6} q_1 - \frac{3+\sqrt{3}}{1} q_2 + \frac{\sqrt{3}}{3} q_3 \\ q_2' &= -\frac{\sqrt{3}+3}{6} q_1 + \frac{3-\sqrt{3}}{1} q_2 + \frac{\sqrt{3}}{3} q_3, \qquad q' = \mathcal{O}(3)q, \\ q_3' &= \frac{\sqrt{3}}{3}(q_1+q_2+q_3), \end{aligned} \tag{13}$$

providing for (12) the form

$$H = \sum_i^3 \frac{1}{2m} p_i'^2 + \frac{3c}{2}(q_1'^2+q_2'^2) + \frac{9k}{2}(q_1'^2+q_2'^2)^2, \tag{14}$$

reducing it to a problem similar to (4).

Let us start with the simplification $k = 0$ and introduce the square root of the 2-body potential

$$w_{ij} = \sqrt{\frac{c}{2}}(q_i-q_j), \quad (w_{12}, w_{23}, w_{31}) \tag{15}$$

With the help of the elements O_{ij} of the matrix (13) the conserved quantities are now

$$J_n=\sum_i (O_{ni}p_i)^2+(O_{ni}w_{ni})^2, \quad n = 1, 2, \tag{16a}$$

$$J_3=\sum_i O_{3i}p_i, \tag{16b}$$

which are obvious and the relation to the Hamiltonian is accordingly to (14) given by

$$H = \frac{1}{2m} J_3^2 + J_1 + J_2. \tag{17}$$

But there exists an additional quantity $(q_1'p_2' - q_2'p_1')$ due to the rotational invariance of (14), providing

$$J_4 = \frac{1}{\sqrt{3}}\left(p_1w_{23} + p_2w_{31} + p_3w_{12}\right). \tag{18}$$

So we have in total 4 independent conserved quantities for the case of 3 harmonically coupled particles, which can also be chosen in the form

$$J_1'= p_1+p_2+p_3 = P; \; J_2'= H, \; J_3'= J_1, \; J_4'= J_4. \tag{19}$$

Now the question is looking at (12) with $k \neq 0$ including the anharmonic coupling. The decomposition (17) now fails, so J_1 and J_2 can no longer be used. But under which conditions is J_4 conserved? $\dot{J}_4 = 0$ demands

$$\frac{\partial V}{\partial q_1}(q_2-q_3)+ \frac{\partial V}{\partial q_2}(q_3-q_1)+ \frac{\partial V}{\partial q_3}(q_1-q_2) = 0, \tag{20a}$$

which is the differential equation for the corresponding admitted potentials. It is seen by inspection, that the potentials r^2 and r^4 obey automaticaly the conditions (20a). It is useful to transform the

equation (20a) with the help of the variables

$$r = q_1- q_2, \; s = q_2- q_3, \; t = q_3- q_1 \text{ with } r + s + t = 0, \tag{20b}$$

in the form of a determinant shape

$$\begin{vmatrix} 1 & 1 & 1 \\ r & s & t \\ V'(r) & V'(s) & V'(t) \end{vmatrix} = 0. \tag{20c}$$

In conclusion we can make the statement that for all potentials obeying (20) with the help of the apropriate variables p_i and w_{ij} all conserved quantities are homogeneous functions of degree $\bar{n} \leq 2$

$$J_n = \sum_{i,j,v} a_{ij}^{nv} \, p_i^{2-v} w_{ij}^{v}, \quad v = 0,1,2; \; j = 1,2,3, \; n = 1 \ldots N \tag{21}$$

where for integrable systems N = f and for "overintegrable" ones N > f.

3.3. Harmonic coupling of f particles.

We will now very shortly sketch the case of f one-dimensional particles on a ring, i.e. with periodic boundary conditions, which are coupled harmonically. The Hamiltonian

$$H = \sum_{i}^{f} \frac{p_i^2}{2m} + \sum_{i,j=1}^{f} \frac{c}{2} (q_i - q_j)^2 \tag{22}$$

can with the help of the orthogonal transformation matrix [3]

$$O = \begin{pmatrix} 1-\alpha & -\alpha & \ldots & -\alpha & \beta \\ -\alpha & 1-\alpha & \ldots & -\alpha & \beta \\ \cdot & \cdot & & & \cdot \\ \cdot & \cdot & & & \cdot \\ \beta & \beta & & & \beta \end{pmatrix} \quad \text{with } 1/\alpha = f - \sqrt{f}, \; 1/\beta = \sqrt{f}, \tag{23a}$$

casted in the form

$$H = \sum_{i}^{f} \frac{1}{2m} p_i'^2 + \sum_{i}^{f-1} \frac{fc}{2} q_i'^2, \tag{23b}$$

representing the center of mass motion and f - 1 uncoupled harmonic oscillators or equivalently a (f - 1) dimensional oscillator. It is again possible to choose as conserved quantities

$$I_n = \sum_{1}^{f} (O_{ni} p_i)^2 + (O_{ni} q_i)^2 \qquad n = 1 \ldots f-1, \tag{24a}$$

$$I_f = \sum_{1}^{f} O_{fi} p_i, \qquad i = 1, \ldots f, \tag{24b}$$

which can also be written as homogeneous functions of the "appropiate variables" p_i and w_{ij} of degree $\bar{n} \leq 2$. Naturally the above system is overintegrable, so there exist more than f conserved quantities. Because the underlying symmetry is SU(f) having a set of (1/2)(f - 1)(f + 2) of commuting generators [2]. there should exist the corresponding number of conserved quantities. But they can not all be independent, because the maximal number should not exceed 2f − 1.

4. Conserved quantities as homogeneous functions

Now we will consider the cases of f one-dimensional particles on a ring with such translationally invariant interaction potentials, that the conserved quantities are besides of the total linear momentum P and the energy H homogeneous functions of the "appropriate variables" p_i and $w_{ij} = \sqrt{V(q_i - q_j)}$ with a degree generally higher than 2. It is advantageous to describe these systems by a f × f matrix of a shape, where the momenta are the diagonal elments. Let us start with the simple example of f = 2 with the matrix

$$M_2(p,w) = \begin{pmatrix} p_i & iw_{12} \\ -iw_{12} & p_2 \end{pmatrix} \tag{25}$$

having the property that the traces

$$I_1 = \mathrm{tr}M = p_1 + p_2 = P,$$
$$I_2 = \tfrac{1}{2}\mathrm{tr}M^2 = \tfrac{1}{2}(p_1^2 + p_2^2) + w_{12}^2 \tag{26}$$

represent the conserved quantities for arbitrary potentials $V(q_1 - q_2)$. The eigenvalues of (25)

$$|M(2) - rI| = 0, \quad r_{1,2} = \frac{P}{2} \pm \sqrt{H_o}, \tag{27}$$

$$H_o = \tfrac{1}{4}(p_1 - p_2)^2 + V(q_1 - q_2),$$

provide another combination of conserved quantities, where H_o is the energy of the relative motion being related to the total energy by

$$H = P^2/4 + H_o.$$

The extension of the above case to f = 3 provides us already with some very instructive result. The matrix

$$M_3(p,w) = \begin{pmatrix} p_1 & iw_{12} & iw_{31} \\ -iw_{12} & p_2 & iw_{23} \\ -iw_{31} & w_{23} & p_3 \end{pmatrix} \tag{28}$$

has the traces

$$J_1 = \mathrm{tr}M = p_1 + p_2 + p_3 = P,$$

$$J_2 = \frac{1}{2} \operatorname{tr} M^2 = \frac{1}{2}(p_1^2 + p_2^2 + p_3^2) + w_{12}^2 + w_{23}^2 + w_{31}^2, \tag{29}$$

$$J_3 = \frac{1}{3} \operatorname{tr} M^3 = \frac{1}{3}(p_1^3 + p_2^3 + p_3^3) + (p_1 + p_2)w_{12}^2 + (p_2 + p_3)w_{23}^2 + (p_3 + p_1)w_{31}^2 \tag{30}$$

where automatically J_1 and J_2 are conserved quantities. Requiring in addition $\dot{J}_3 = 0$ gives us a constraint on the potentials of a type similar to (20b)

$$\frac{\partial V_{12}}{\partial q_1}(V_{23} - V_{31}) + \frac{\partial V_{23}}{\partial q_2}(V_{31} - V_{12}) + \frac{\partial V}{\partial q_3}(V_{12} - V_{23}) = 0. \tag{31}$$

Again we introduce the variables

$$r = q_1 - q_2, \quad s = q_2 - q_3, \quad t = q_3 - q_1 \text{ with } r + s + t = 0, \tag{32}$$

providing for condition (31) a determinant analogous to (20c)

$$\dot{J}_3 = 0 \quad \begin{vmatrix} 1 & 1 & 1 \\ V(r) & V(s) & V(t) \\ V'(r) & V'(s) & V'(t) \end{vmatrix} = 0 \quad \text{with } r + s + t = 0. \tag{33}$$

So all potentials obeying the condition (33) lead to integrable systems for $f = 3$ and the conserved quantities are given by (29) and (30). In the case of $f = 2$ we found that all (positiv) potentials $V(q_k - q_l)$ lead to integrable systems, in the case of $f = 3$ only those fullfilling (33) are permitted, the question is now how an extension to $f \geq 4$ would look like. A surprising fact is observed. Replacing (31) by

$$\frac{\partial V_{ij}}{\partial q_i}(V_{jk} - V_{ki}) + \frac{\partial V_{jk}}{\partial q_j}(V_{ki} - V_{ij}) + \frac{\partial V_{ki}}{\partial q_k}(V_{ij} - V_{jk}) = 0, \tag{34}$$

$$i, j, k = 1 \ldots f.$$

we get $\binom{f}{3}$ conditions. If they are fullfilled the extension of the matrix (28) to arbitary f leads to conserved quantities

$$J_n = \frac{1}{n} \operatorname{tr} M^n, \quad \dot{J}_n = 0, \; n = 1 \ldots f. \tag{35}$$

It is now a rather tedious task to look for the solutions of equation (33) involving the addition theorem of the potential, i.e. expressing $V(r + s)$ by $V(r)$ and $V(s)$. So far the following solutions are known [4]:

i. in the class of ration al functions:

$$V(r) \sim r^{-2}$$

ii. in the class of exponential functions:

$V(r) \sim e^{-r}$,

iii. in the class of simply periodic functions:

$V(r) \sim \sin^{-2} r,\ \cot^{-2} r,\ \sinh^{-2} r,\ \coth^{-2} r,$

iv. in the class of doubly peridic functions

$V(r) \sim sn^{-2}(r),\ cn\,(r)/sn^2(r),\ dn\,(r)/sn^2(r),\ dn^2(r)/sn^2(r),$

$cn^2(r)/sn^2(r)dn^2(r),\ cn^2dn^2(r)/sn^2(r)$

and $dn^2(r)/sn^2(r)cn^2(r)$,

where $sn(r,k)$, $cn(r,k)$ and $dn(r,k)$ are Jacobison elliptic functions [5] with modul $k \leq 1$.

The doubly periodic Weierstrass-function [5] of the complex variable r with the periods 2w and 2w' solves generally the above differential condition (33), giving a general solution of the form

$V(r) \sim \wp(r,w,w')$,

indicating that all potentials fulfilling the addition theorem of the $\wp$-function lead to integrable systems. Special versions are just the ones given by iv. being expressed by the Jacobian elliptic functions.

4. Concluding remarks

It is the aim of these considerations to demonstrate that in the case of f one-dimensional particles on a ring, i.e. with periodic boundary conditions $q_{f+1} = q_1$, for a wide class of translationally invariant potentials, which are charaterised by the two determinants (20c) and (33), the many-particle systems are integrable and the conserved quantities can in principle be calculated explicitly. Many details of the results can be obtained by different methods [6], but the one presented here seems to be rather clear and straightforward, giving the explicit result, that all the corresponding conserved quantities are homogeneous functions of the appropiate variables p_i and w_{ij}. For the case (20c) the degree does not exceed 2, while for (33) the degree varies from 1 ... f. This result has so far not been stated elsewhere at least to the knowledge of the author. But this is only a special class of integrable many-particles systems varying from harmonic to anharmonic conpling, from the exponential coupling to the hard core potential r^{-2} and all simply and doubly periodic potentials with the same quadratic singularily at the origin. It remains to investigate the question, if it is possible to find completely other structures of conserved quantities, perhaps expressed with different variables and which are the connected potentials.

The considered method is limited so far only to f one-dimensional particles. It is a question up to now, if there exists a way to extend it to the treatment of f three-dimensional particles. Surely in these systems the conserved quantities will have a more complicated structure. But severall attempts are prepared to discuss also these systems and to relate the results to the treatment of nonlinear evolution equations in (3+1)-dimensions [7].

References

[1] Thirring, W.: Lehrbuch der Mathm. Physik, Bd.1: Klassische Dynamische Systeme. Springer-Verlag, Wien - New York, 1977,
Arnold, V.J.: Mathematische Methoden der Klassischen Mechanik, VEB Deutscher Verlag der Wissenschaften, Berlin 1988.

[2] Gell-Mann, M. Ne'eman, Y.: The Eightfold Way, W.A. Benjamin, Inc, New York - Amsterdam 1964.

[3] Moebius, P.: Nuclear Physics 16(1960)278

[4] Calogero, F.: Lettere Nuovo Cimento 16(1976)7780.
Kunick, A., Steeb, W. H.: Chaos in dynamischen Systemen, BI Wissenschaftsverlag, Mannheim (Wien) Zuerich 1986.

[5] Jahnke, E., Emde, F.: Tafeln hoeherer Funktionen, B. G. Teubner Verlagsgesellschaft, Leipzig 1948.

[6] Olshanetsky, M.A., Perelomov, A.M.; Physics Reports 71(1981)313
Tachtadzjan, L. A., Faddeev, L.D.: Hamiltonov podchod v teorii solitonov. Moskva, Nauka, 1986

[7] Moebius, P.,, to be published.

K-AUTOMORPHISMS IN QUANTUM THEORY

Heide Narnhofer[1]

Abstract

Two possibilities to define quantum K-systems are offered and their consequences are compared. Examples are given and open problems are pointed out.

1 Introduction

In the context of measure theory the approach to equilibrium is well understood and described in the theory of dynamical systems. We can distinguish between different mixing properties and know that the best mixing occurs for K-systems. Here a quantitative measure for mixing is offered by the dynamical entropy. In addition we know that in symbolic dynamics K-systems correspond to Bernoulli shifts so that we can essentially classify the possible K-systems.

We know that the quantum analogue to a measure space are the von Neumann algebras. Here the dynamics is given by an automorphism group α_t ($t \in \mathbf{Z}$ or $t \in \mathbf{R}$). So it is natural to construct an ergodic theory also for von Neumann algebras. For a long time quantum ergodic theory concentrated on the problem how an invariant state can be decomposed into invariant ones [1]. We know that the decomposition is unique if the automorphism is weakly asymptotically abelian (what is trivially satisfied for an abelian von Neumann algebra).

Starting with an extremal invariant state on a von Neumann algebra, we call this state ergodic and can study its mixing properties. The correspondence to the classical theory is fairly evident [2]. To translate K-mixing turns out to be less trivial, but there exist two different successful approaches. Unfortunately, it is still an open problem whether the two approaches coincide. One has the advantage to offer the possibility to examine whether an automorphism is a K-automorphism, the other to construct a K-automorphism.

In the following we will give the two definitions, point out some consequences and some relations, and finally report on the examples.

2 Entropic K-Systems

Definition: [3] Let $(\mathcal{M}, \alpha, \omega)$ be a von Neumann algebra with α an automorphism and ω an α-invariant state. Let $\mathcal{A}$ be a finite-dimensional subalgebra. Then $(\mathcal{M}, \alpha, \omega)$ is an entropic K-system, if

$$\lim_{n\to\infty} h_\omega(\alpha^n, \mathcal{A}) = H_\omega(\mathcal{A}) \qquad \forall\, \mathcal{A}, \tag{1}$$

[1]Institut für Theoretische Physik, Universität Wien, Boltzmanngasse 5, A-1090 Wien

where $H_\omega(\mathcal{A})$ is the entropy of $\mathcal{A}$ with respect to $(\omega, \mathcal{A})$ and $h_\omega(\alpha^n, \mathcal{A})$ the dynamical entropy of α^n with respect to $\mathcal{A}$.

The definitions for $H_\omega(\mathcal{A})$ and $h_\omega(\alpha^n, \mathcal{A})$ can be found in [4] in this volume, respectively are given in [5,6].

Remark: For classical systems usually (1) is replaced by the condition [7,8]

$$h_\omega(\alpha, \mathcal{A}) > 0 \qquad \forall \mathcal{A}. \tag{2}$$

In [3] it is shown that in this case (2) implies (1). In the quantum case we do not have a proof nor do we know a counterexample.

We have immediately the consequences

Proposition: Let $(\mathcal{M}, \alpha, \omega)$ be an entropic K-system. Then ω is extremal invariant.

Proof: If there exists another invariant state $\bar{\omega}$, then $\bar{\omega}(A) = \omega(BA)$ with $B \in \mathcal{M}$ and $\alpha B = B$. Then $h_\omega(\alpha^n, \{B\}'') \equiv 0$.

Proposition: There does not exist any quasiperiodic subalgebra $\mathcal{A} \subset \mathcal{M}$, such that $\alpha^n \mathcal{A} \overset{\delta_n}{\subset} \mathcal{A}$ with $\lim \delta_n = 0$. If α is implemented by U, then the pure point spectrum of U apart from the nondegenerate 1 (corresponding to the invariant state ω) must therefore be empty.

Proof: $h_\omega(\alpha^n, \mathcal{A}) < \delta_n \ln \dim \mathcal{A}$.

Example 1: Let $\mathcal{M}$ be the von Neumann algebra corresponding to the even Fermi algebra built up by $a(f)$, $a^\dagger(f)$. Let α be the automorphism given by $\alpha a(f) = a(Uf)$ and assume U has an absolutely continuous spectrum. Then $\mathcal{M}$ is an entropic K-system.

The proof relates the automorphism to a tensor product of shifts, then uses the fact that the shift gives rise to an algebraic K-system and is strongly asymptotically abelian (see next chapter).

Example 2: Take the complete Fermi algebra of the above example. Then $\mathcal{M}$ is not an entropic K-system but

$$h_\omega(\alpha^n, \{a^\dagger + a + 1\}'') \leq \frac{1}{2} H_\omega(\{a^\dagger + a + 1\}'').$$

It is an open question whether it is in fact 0 (we do not know whether (2) $\rightarrow$ (1)).

Example 3: Take again Example 1, but now assume U has pure point spectrum. Then

$$h_\omega(\alpha^n, \mathcal{A}) \equiv 0.$$

Example 4: Assume U has singular continuous spectrum. This example is under investigation, but we conjecture that again $h_\omega(\alpha^n, \mathcal{A}) = 0$.

The idea in the definition (1) is that $\alpha^n \mathcal{A}$ and $\mathcal{A}$ become completely independent for increasing n. The idea is supported by the following result:

Theorem: [9] Assume $(\mathcal{M}, \alpha, \omega)$ is an entropic K-system and $\mathcal{M}$ type II_1 factor and ω the trace. (It must be the trace, because the trace is invariant and is by assumption the only invariant state.) Then

$$\text{st-}\lim[A, \alpha^n B] = 0 \qquad \forall A, B,$$

so α is strongly asymptotically abelian.

Proof: The proof is based heavily on the definition of h and H, which for the trace can be calculated explicitly. We show that

$$\lim \omega(P\tau^n Q) - \omega(P)\omega(Q) = 0$$

for projections which gives weak asymptotic abelianess and, in addition, that $P\tau^n Q \geq -\varepsilon$ for large n, which gives strong asymptotic abelianess. For arbitrary ω we believe that the result remains true but the proof must be based on new ideas. Hopefully, also in the quantum case a kind of symbolic dynamics should be useful.

A trivial corollary is

Corollary: The system is strongly mixing (hyperclustering) [2], i.e.

$$\lim \omega(A\tau^n B \cdot C\tau^n D) = \omega(AC)\omega(BD).$$

Remark: For the classical situation strongly mixing and weakly mixing, i.e. $\lim \omega(A\tau^n B) = \omega(A)\omega(B)$ are equivalent. Here mixing is strengthened to K-mixing, so that mixing for chosen A is in an appropriate way uniform in B. This uniformity can be expressed only in the language of algebraic K-systems. But also in the quantum case we have not used all information coming from the fact that we are dealing with an entropic K-system. A candidate to examine is

$$\lim_{n \to \infty} \frac{\omega(P\tau^n P \ldots \tau^{nk} P)}{\omega(P)^k} = 1$$

since this expression enters in the calculation of $h(\alpha^n, \{P\}'')$. So far we do not know its convergence properties with respect to varying k even in the classical situation.

The reason why we try to improve the mixing properties comes from the following observation: In classical theory we know that every K-system has infinite Lebesgue spectrum, so in particular its spectrum apart from eigenvalue 1 is absolutely continuous. Strong asymptotic abelianess still allows singular continuous spectrum but it seems doubtful that quantum theory should be less restrictive (compare Example 4).

3 Algebraic K-Systems

Definition: $(\mathcal{M}, \alpha, \omega)$ is called an algebraic K-system, if $\omega = \omega \circ \alpha$, and there exists a subalgebra $\mathcal{M}_0$, such that [10,11]

(i) $\alpha \mathcal{M}_0 \supset \mathcal{M}_0$,

(ii) $\bigwedge_{n=0}^{-\infty} \alpha^n \mathcal{M}_0 = \lambda \mathbf{1}$,

(iii) $\bigvee_{n=0}^{\infty} \alpha^n \mathcal{M}_0 = \mathcal{M}$.

The intersection is understood as intersection of sets. The union is understood as the algebra generated by its constituents.

Theorem: Let $\mathcal{M}$ be abelian. Then $(\mathcal{M}, \alpha, \omega)$ is an algebraic K-system, iff it is an entropic K-system [8].

The proof is based on the fact that every finite-dimensional subalgebra has a trivial tail, i.e.

$$\bigwedge_{n=0}^{\infty} \bigvee_{n'=0}^{\infty} \alpha^{-n-n'} \mathcal{B} = \lambda \mathbf{1}.$$

This statement does not hold for a nonabelian algebra because we lose the control over $\bigvee_{n'=0}^{\infty} \alpha^{-n'} \mathcal{B}$. Examine the shift for the continuous Fermi system

$$\alpha_x a(f) = a(f_x) \quad \text{with} \quad f_x(y) = f(y - x).$$

Take $f(x) = e^{-x^2}$. Then $\int \bar{g}(x) e^{-(x+y)^2} dx \equiv 0 \ \forall y \leq 0$ implies $g \equiv 0$. So $\bigvee_{x=0}^{-\infty} \alpha_x \{a(f), a(f)^\dagger\}'' \equiv \mathcal{M}$, though we know that with $\bar{f}(x) = f(x)\Theta(-x)$ we have an example for an algebraic K-system [12].

Let us first collect consequences:

Proposition: An algebraic K-system is ergodic, i.e. ω is the only invariant state.

Proof: Let $m = \alpha m$ be an invariant element. Then due to (iii) $m \overset{\delta}{\in} \alpha^n \mathcal{M}_0$ for some n and also $\alpha^k m \overset{\delta}{\in} \alpha^n \mathcal{M}_0 \; \forall k$, i.e. $m \overset{\delta}{\in} \bigwedge_{n=0}^{-\infty} \alpha^n \mathcal{M}_0 = \lambda \mathbf{1}$. So $m = \lambda \mathbf{1}$.

Proposition: The system is K-clustering, i.e. for every $m \in \mathcal{M} \; \exists n$ such that

$$|\omega(m\alpha^{-n}\bar{m}) - \omega(m)\omega(\bar{m})| \leq \varepsilon\omega(\bar{m}) \qquad \forall \bar{m} \in \mathcal{M}_0.$$

Remark: Take $\bar{m} = m_1 \alpha^{-1} m_1 \ldots \alpha^{-k} m_1$ and compare with our remark on entropic K-systems.

Proposition: Let U implement α. Then U has infinite Lebesgue spectrum [11].

Proof: Define the projection P_0 by $P_0 \mathcal{H} = \overline{\mathcal{M}_0 \Omega}$. Then $U^n P_0 U^{\dagger n} = P_n$ with $P_n \geq P_0$. Therefore we have a kind of Weyl relations for U and P_n which gives the spectral properties of U.

Remark: For entropic K-systems we have been looking for a similar property, i.e. whether U has absolutely continuous spectrum and this spectrum is R and infinitely degenerate (the last would be a consequence of asymptotic abelianess).

Proposition: Let $(\mathcal{M}, \alpha, \omega)$ be an algebraic K-system. Then α is weakly asymptotically abelian.

Proof: This is a consequence of K-clustering.

Proposition: Let $(\mathcal{M}, \alpha, \omega)$ be an algebraic K-system. Let α be strongly asymptotically abelian. Then $(\mathcal{M}, \alpha, \omega)$ is an entropic K-system.

For the proof see [13].

Proposition: Let $(\mathcal{M}, \alpha, \omega)$ be an algebraic K-system. Take φ some state over $\mathcal{M}$. Then the relative entropy satisfies

$$\begin{aligned} \lim_{n\to\infty} (S(\omega \circ \alpha^{-n} | \varphi \circ \alpha^{-n})_{\mathcal{M}_0} &= S(\omega|\varphi) \\ \lim_{n\to\infty} S(\omega \circ \alpha^{+n} | \varphi \circ \alpha^{+n})_{\mathcal{M}_0} &= 0. \end{aligned}$$

These limits are attained monotonically [14].

Proof:

$$S(\omega \circ \alpha^n | \varphi \circ \alpha^n)_{\mathcal{M}_0} = S(\omega|\varphi)_{\alpha^{-n}\mathcal{M}_0} \overset{n>0}{\leq} S(\omega|\varphi)_{\mathcal{M}_0}$$

and

$$\lim \|\omega_{|\mathcal{M}_0} - \varphi \circ \alpha^n{}_{|\mathcal{M}_0}\| = 0.$$

The main problem that we see is to decide for a given system $(\mathcal{M}, \alpha, \omega)$ whether there exists some $\mathcal{M}_0$ that makes it an algebraic K-system. For a classical system we can start with any finite-dimensional algebra and construct $\mathcal{N}_0 = \bigvee_{n=0}^{\infty} \alpha^{-n}\mathcal{B}$. Then it can happen that $\bigvee_{n=0}^{\infty} \alpha^n \mathcal{N}_0$ is smaller than $\mathcal{M}$, so we have to enlarge $\mathcal{B}$ but finally we will succeed to get everything. Evidently, $\mathcal{M}_0$ is not unique nor do we have an order relation between different $\mathcal{M}_0$ and $\bar{\mathcal{M}}_0$. This method fails in the quantum case. Therefore we have to start with the spectral property of U, find P_0 (which by far is not unique since U is infinitely degenerate) and check whether some P_0 corresponds to a subalgebra by $P_0\mathcal{H} = \overline{\mathcal{M}_0\Omega}$.

Another problem that is, of course, related to the above is the following. Given two algebraic K-systems $(\mathcal{M}, \alpha, \omega, \mathcal{M}_0)$ and $(\mathcal{M}, \bar{\alpha}, \bar{\omega}, \bar{\mathcal{M}}_0)$, when are they equivalent, i.e. does there exist a γ such that $\omega \circ \gamma = \bar{\omega}$ and $\gamma^{-1} \circ \alpha \circ \gamma = \bar{\alpha}$. (Compare in classical theory: two Bernoulli shifts are equivalent, if they have the same dynamical entropy [8].)

4 Examples of Algebraic K-Systems

There exist, of course, the simple generalization of the Bernoulli shift:

Example: Take $\mathcal{M} = \bigotimes_{n=-\infty}^{+\infty} M_k^{(n)}$ with M_k a $k \times k$ matrix and α the shift and ω an α-invariant state, e.g. the trace. With $\mathcal{M}_0 = \bigotimes_{n=0}^{\infty} M_k^{(n)}$ $(\mathcal{M}, \alpha, \omega, \mathcal{M}_0)$ is an algebraic K-system.

Of course, M_k can be taken to be some subalgebra of the $k \times k$ matrices, so variation of the construction seems possible. But, in fact, the structure is much richer. It is based on the results of [15].

Theorem: [15] Let $\mathcal{M}_0 \supset \mathcal{M}_{-1}$ be two factors of type II_1 (resp. type III [16]). Then it is possible to construct a chain

$$\ldots \mathcal{M}_1 \supset \mathcal{M}_0 \supset \mathcal{M}_{-1} \supset \mathcal{M}_{-2} \ldots$$

such that the index $[\mathcal{M}_i : \mathcal{M}_{i-1}] = \lambda^{-1}$. The chain contains a subalgebra $\mathcal{R}_\lambda$ such that

$$\begin{array}{ccccccc} \ldots & \mathcal{M}_1 & \supset & \mathcal{M}_0 & \supset & \mathcal{M}_{-1} & \ldots \\ & \cup & & \cup & & \cup & \\ \ldots & \mathcal{R}_{\lambda,1} & \supset & \mathcal{R}_{\lambda,0} & \supset & \mathcal{R}_{\lambda,-1} & \ldots \end{array}$$

with $\mathcal{R}_{\lambda,k} = \{e_k, e_{k-1}, \ldots,\}''$ where

$$e_k e_{k+i} e_k = \lambda e_k \qquad |i| = 1$$

and

$$e_k e_{k+1} = e_{k+i} e_k \qquad |i| \geq 2$$

and e_k is connected with conditional expectations E_{k-1} from $\mathcal{M}_k$ to $\mathcal{M}_{k-1}$ by

$$e_k m_{k-1} e_k = E_{k-2}(m_{k-1}) e_k, \qquad m_{k-1} \in \mathcal{M}_{k-1}$$

and $\mathcal{M}_k = \{\mathcal{M}_{k-1}, e_k\}''$.

We have to relate this chain to automorphisms.

Theorem: Let $\mathcal{M}_0$ be a type III algebra and $\mathcal{M}_{-1} \subset \mathcal{M}_0$ is isomorphic to $\mathcal{M}_0$ (and so $\mathcal{M}_{-2}$). Choose Ω that is cyclic and separating for $\mathcal{M}_0$ and $\mathcal{M}_{-1}$ with corresponding conjugate unitary J_0 and J_{-1}. Define

$$\gamma m = J_{-1} J_0 m J_0 J_{-1}.$$

Then γ is an isomorphism from $\mathcal{M}_{2k} \to \mathcal{M}_{2(k-1)}$ [17].

The same construction is possible for type II algebras [18] though here, due to the lack of an appropriate Ω, more care has to be taken in the definition of J_{-1} and J_0. This isomorphism is called canonical shift and is in fact an automorphism for $\mathcal{M} = \bigcup_{k=0}^{\infty} \mathcal{M}_k$.

Theorem: [14] Let $\mathcal{M}_0$ be type II_1 and $[\mathcal{M}_0 : \mathcal{M}_{-1}]$ have finite index. Let α be an isomorphism from $\mathcal{M}_0$ to $\mathcal{M}_{-1}$. Then the construction of [15] can be done in such a way that

$$\alpha^n \mathcal{M}_0 = \mathcal{M}_{-n}$$

and $\alpha^n e_k = e_{k-n}$.

This should be compared with

Theorem: [19] If $[\mathcal{M}_0 : \mathcal{M}_{-1}]$ is infinite, then it is still possible to construct the infinite chain and also the automorphism γ from $\mathcal{M}_0$ to $\mathcal{M}_{-2}$, because the imbedding $\mathcal{M}_0 \supset \mathcal{M}_{-1}$ and $\mathcal{M}_{-1} \supset \mathcal{M}_{-2}$ are dual to one another but not selfdual.

Since we are interested in algebraic K-systems we wonder whether the above constructions offer examples. The imbedding relations are satisfied, it remains to check whether the tail $\bigcap_{n=0}^{\infty} \mathcal{M}_{-n} = \lambda \mathbf{1}$.

Remark: Let $\mathcal{M}_0 \supset \mathcal{M}_{-1}$. Then $\mathcal{M}_0 \otimes M_k \approx \mathcal{M}_0$ and so is $\mathcal{M}_{-1} \otimes \mathcal{M}_k$. Then $\bigcap \mathcal{M}_{-n} \otimes M_k \supset M_k$. Therefore we cannot check from the imbedding without additional assumptions that γ is ergodic. We have to remove the freedom in the choice of the chain.

Theorem: Let $\mathcal{M}_0 \supset \mathcal{M}_{-1}$ be enlarged to two chains

$$\begin{array}{ccccccccc} \cdots \supset & \mathcal{M}_1 & \supset & \mathcal{M}_0 & \supset & \mathcal{M}_{-1} & \supset & \mathcal{M}_{-2} & \cdots \\ & & \supset & \mathcal{M}_0 & \supset & \mathcal{M}_{-1} & \supset & \bar{\mathcal{M}}_{-2} & \cdots \end{array}$$

Define $\mathcal{A}_k = \mathcal{M}_0 \cap \mathcal{M}'_{-k}$ resp. $\bar{\mathcal{A}}_k = \mathcal{M}_0 \cap \bar{\mathcal{M}}'_{-k}$. Then $\mathcal{A}_k$ and $\bar{\mathcal{A}}_k$ are isomorphic. Therefore

Corollary: γ is ergodic and $(\mathcal{M}, \gamma, \omega)$ an algebraic K-system, iff

$$\bigcup_{k=0}^{\infty} \mathcal{A}_k = \mathcal{M}_0.$$

There exists a

Theorem [20,21]: Let k_j be the cardinal number of simple summands in $\mathcal{M}_0 \cap \mathcal{M}'_{-j}$. If $\sup(k_j) < \infty$, then there exists a choice for the chain $\{\mathcal{M}_n\}$ such that

$$\bigcup_{k=0}^{\infty} \mathcal{A}_k = \mathcal{M}_0$$

or equivalently a choice of γ, such that $(\mathcal{M}, \gamma, \omega)$ is an algebraic K-system and γ is weakly asymptotically abelian.

It is not possible to choose the chain $\{\mathcal{M}_n\}$ also in such a way that γ is strongly asymptotically abelian. In fact, γ is then determined by the $\mathcal{A}_k$ and therefore no freedom is left to adjust strong asymptotic abelianess. But we have

Theorem: Let $(\mathcal{M}, \alpha, \omega)$ be an algebraic K-system. Then there exists a subalgebra $\mathcal{N} \subset \mathcal{M}$, such that $(\mathcal{N}, \alpha, \omega)$ is an entropic K-system.

Proof: $\mathcal{R}_\lambda$ serves the purpose.

We remain with one deep problem: so far we have concentrated on an automorphism group α^n, $n \in \mathbf{Z}$. But in the physical situation α should define time evolution, so α_t is a continuous group, $t \in \mathbf{R}$. Concerning the entropic K-system no problem arises. We have to consider α_{kn} with k some number and use continuity and asymptotic additivity of the entropy. So $(\mathcal{M}, \alpha_t, \omega)$ is an entropic K-system for all t, if it is for $(\mathcal{M}, \alpha_1, \omega)$.

For the algebraic K-system the situation changes drastically. Assume $(\mathcal{M}, \alpha^2, \omega)$ is an algebraic and entropic K-system. Thus also $(\mathcal{M}, \alpha, \omega)$ is an entropic K-system (we are considering limit behaviour of α^n). But so far we have no proof whether $(\mathcal{M}, \alpha, \omega)$ is an algebraic K-system. In fact, given $\mathcal{M}_0$

that constructs $\mathcal{M}$ with respect to α^2, so that $[\mathcal{M}_0 : \alpha^{-2}\mathcal{M}_0] < 4$, then we know that it is impossible to imbed some $\alpha^{-1}\mathcal{M}_0$ between $\mathcal{M}_0$ and $\alpha^{-2}\mathcal{M}_0$. But the same is true for a Bernoulli shift of order 2. All we have to do here is to change to a Bernoulli shift of order 4 with the same dynamical entropy. Then the two are isomorphic, but the Bernoulli shift of order 4 can trivially be decomposed into two of order 2. So we have to look for a similar construction also in the quantal situation.

References

[1] D. Ruelle, Statistical Mechanics, Benjamin, Amsterdam 1969.

[2] H. Narnhofer, W. Thirring, Mixing Properties of Quantum Systems, to be published in Journ. Stat. Phys.

[3] H. Narnhofer, W. Thirring, Quantum K-Systems, to be published in Commun. Math. Phys.

[4] T. Hudetz, Algebraic Topological Entropy, Vienna preprint UWThPh-1989-52, these proceedings.

[5] A. Connes, H. Narnhofer, W. Thirring, Commun. Math. Phys. 112 (1987) 691.

[6] A. Connes, C.R. Acad. Sci. Paris t 301 (1985) 1.

[7] P. Walters, An Introduction to Ergodic Theory, Springer, New York 1982.

[8] P. Cornfeld, S.V. Fomin, Ya.G. Sinai, Ergodic Theory, Springer, Berlin 1982.

[9] F. Benatti, H. Narnhofer, Asymptotic Abelianess, to be published in Commun. Math. Phys.

[10] G.G. Emch, Commun. Math. Phys. 49 (1976) 191.

[11] W. Schröder, in: Quantum Probability and Application to the Theory of Irreversible Processes, p. 340, L. Accardi, A. Gorini, ed., Springer, Berlin 1984.

[12] H. Narnhofer, A. Pflug, W. Thirring, Mixing and Entropy Increase in Quantum Systems, Vienna preprint UWThPh-1989-27.

[13] H. Narnhofer, Dynamical Entropy, Quantum K-Systems and Clustering, to be published in Quantum Probability V, L. Accardi, W. von Waldenfeld, ed.

[14] H. Narnhofer, W. Thirring, Algebraic K-Systems, Vienna preprint UWThPh-1989-17.

[15] V.F.R. Jones, Invent. Math. 72 (1983) 1.

[16] H. Kosaki, J. Funct. Anal. 66 (1986) 123.

[17] R. Longo, Index of Subfactors and Statistics of Quantum Fields II, Roma preprint 1989.

[18] A. Ocneanu, Quantized groups, string algebras and Galois theory, preprint.

[19] R. Herman, Talk at the Confernece on Operator Algebras and Quantum Groups in Cortona, 1989.

[20] M. Choda, Entropy for Canonical Shifts, Univ. of Osaka preprint 1989.

[21] S. Popa, Classification of subfactors, the reduction to commuting squares, University of California preprint.

CELLULAR AUTOMATA AND SYMBOLIC DYNAMICS

Uwe Quasthoff and Axel Schüler[1]

1. Introduction

In this paper we consider onesided, onedimensional cellular automata[2] (CA), i.e. CA acting on a lattice isomorphic to $\mathbb{N}$. Assume that each automaton has a state space $S=\{0,1,\dots,p-1\}$ equipped with the discrete topology. Then the configuration space $S^{\mathbb{N}}$ of all possible states $s=(s_1, s_2, s_3, \dots)$ ($s_k \in S$) is the cartesian product of countably many copies of S with the product topology. A metric d on $S^{\mathbb{N}}$ can be given by

$$d(s,s') = \sum_{k=1}^{\infty} \frac{|s_k - s'_k|}{2^k} .$$

$s_1, s_2, s_3, \dots$ are called the co-ordinates of s.

A natural and much studied transformation on S is the so-called left shift σ defined by

$$\sigma(s) = s' \quad \text{with } s'_k = s_{k+1},$$

see, for instance [2,3].

A local transition rule of a CA is a function $f: S \to S^n$ for some $n \geq 2$. We will always assume that the function $f(x_1, x_2, \dots, x_n)$ explicitely depends on its arguments x_1 and x_n. The set of all those local rules will be denoted by F(n). Then each local transition rule f gives rise to a global transition rule $\tilde{f}$ defined by

$$(\tilde{f}(s))_k = f(s_k, s_{k+1}, \dots, s_{k+n-1}). \qquad (1)$$

[1]Karl-Marx-Universität Leipzig, Sektion Mathematik, DDR-7010 Leipzig

[2]The corollary 5.1 deals with twosided CA. In this case in the following definitions we have to replace $\mathbb{N}$ by $\mathbb{Z}$.

On the other hand, every global transition rule of a onesided CA has the form $\sigma^m \circ \tilde{f}$ for some local rule f and some natural number m. In the case of a twosided CA n can be an arbitrary integer.

2. Vertical shifts induced by cellular automata

As a connection between a transition function f an a shift we consider for a fixed natural number m the following mapping ψ_m which projects at the first m states of global configuration:

$$\psi_m((s_1, s_2, \ldots, s_m, s_{m+1}, \ldots)) = (s_1, s_2, \ldots, s_m).$$

Next for a given global transition rule f of a CA, ψ_m induces a mapping

$$(\Phi_m s)_k = \psi_m(f^k(s)). \qquad (2)$$

Example 1. Illustration of the mapping Φ_2 for S={0,1} and the rule number 90.

The local rule f for the rule number 90 (for a definition of the numbering of the rules, see, for instance, [4]) is given by $f(x_1, x_2, x_3) = x_1 + x_3 \pmod 2$. If we choose an arbitrary state s and print the states f(s), $f^2(s)$, $f^3(s)$,... in consecutive lines we get

```
0 0 1 0 1 1 0 1 0 1 1 0 0 1 0 1 . .
1 0 0 1 1 0 0 0 1 1 1 1 0 0 . . . .
1 1 1 1 1 0 1 1 0 0 1 1 . . . . . .
0 0 0 1 0 1 1 1 1 1 . . . . . . . .
0 1 0 0 1 0 0 0 . . . . . . . . . .
0 1 1 0 1 0 . . . . . . . . . . . .
1 1 0 0 . . . . . . . . . . . . . .
1 1 . . . . . . . . . . . . . . . .
. . . . . . . . . . . . . . . . . .
```

Let us call this the graph of the CA for a given initial state s. In general Φ takes as argument the sequence in first line and its value is the sequence formed by the m-tuples in the first m rows. Roughly speaking Φ transforms a horizontal into a vertical sequence.

Proposition 2. The mapping $\Phi_m : S \to (S^m)$ described above is continuous with respect to the product topologies of the discrete topologies on S and S^m, respectively.

Proof. This follows directly from the continuity of $\tilde{f}$. ■

Proposition 3. Let σ' be the (vertical) shift on $(S^m)^{\mathbb{N}}$ and f and Φ_m as above. Then

$$\sigma' \circ \Phi_m = \Phi_m \circ f.$$

Proof. This is clear from the above example. ■

3. Topological conjugacy

Now we are interested in the case in which the transformation Φ_m is invertable and Φ_m^{-1} is again continuous. Recall that two transformations f and g on topological spaces X and Y, respectively, are called topologically conjugate if there is a homeomorphism $\Psi: X \to Y$ such that $\Psi \circ f = g \circ \Psi$.

Definition 4. The local rule $f \in F(n)$ is called left permuting if for all fixed $(n-1)$-tuples $(\tilde{x}_2, \ldots, \tilde{x}_n) \in S^{n-1}$ the function $f(x_1, \tilde{x}_2, \ldots, \tilde{x}_n)$ considered as a function of x_1 is a permutation. Similarily we define right permuting rules.

Theorem 5. Let the local rule $f \in F(n)$ be right permuting. Then the onesided CA with the global rule $\tilde{f}$ corresponding to f is topologically conjugate to the onesided p^{n-1}-shift.

Proof. We will show that the mapping Φ_2 defined in (2) is the disired homeomorphism. We already proved the continuity of Φ_2. By the compactness of $S^{\mathbb{N}}$ we only have to show that Φ_2 is one to one.
Consider the following part of a graph of a CA (superscripts refer to

line numbers):

$$\begin{matrix} x^t_k & x^t_{k+1} & \ldots & x^t_{k+n-2} & x^t_{k+n-1} \\ x^{t+1}_k & & & & \end{matrix}$$

Let us fix $x^t_k, x^t_{k+1}, \ldots, x^t_{k+n-2}$ for the moment. Now f is right permuting, hence there is a 1-1 correspondence between the values for x^t_{k+n-1} and x^{t+1}_k. This implies the bijectivity of Φ_2.

■

The following example shows the explicit construction of the inverse of Φ_2.

<u>Example 6.</u> Φ_2^{-1} for rule 150.

The local rule is given by $f(x_1, x_2, x_3) = x_1 + x_2 + x_3 \pmod 2$.

Hence, four elements of the graph positioned as

$$\begin{matrix} w & x & y \\ z & & \end{matrix}$$

always satisfy $w+x+y+z=0 \pmod 2$. So given a vertical sequence of two columns (we have $f \in F(3)$) we can use Φ_2^{-1} to reconstruct the upper right part of the graph step by step as follows.

```
. . . . . . .     . . . . . . .     . . . . . . .
0 1 . . . . .     0 1 . . . . .     0 1 . . . . .
1 1 . . . . .     1 1 1 . . . .     1 1 1 1 . . .
1 1 . . . . .     1 1 . . . . .     1 1 . . . . .
      a)                b)                c)

. . . . . . .     . . . . . . .     . . . . . . .
0 1 0 . . . .     0 1 0 0 . . .     0 1 0 0 1 . .
1 1 1 1 . . .     1 1 1 1 . . .     1 1 1 1 . . .
1 1 . . . . .     1 1 . . . . .     1 1 . . . . .
      d)                e)                f)

. . . . . . .
0 1 0 0 1 0 .
1 1 1 1 . . .
1 1 . . . . .
      g)
```

Using the first two columns of the graph of example 1 and the above construction, we get

```
0 0 1 1 0 0 1 0 0 1 1 0 0 0 0 1 . .
1 0 0 1 1 1 1 1 0 0 1 0 0 1 . . . .
1 1 0 1 1 1 0 1 1 1 1 1 . . . . . .
0 0 0 1 0 0 0 1 1 1 . . . . . . . .
0 1 1 1 0 1 0 1 . . . . . . . . . .
0 1 0 0 1 0 . . . . . . . . . . . .
1 1 1 1 . . . . . . . . . . . . . .
1 1 . . . . . . . . . . . . . . . .
. . . . . . . . . . . . . . . . . .
```

The next corollary gives an application of the above theorem to twosided CA.

<u>Corollary 5.1.</u> Let the local rule $f \in F(n)$ be both left an right permuting. Then for $0<k<n-1$ the twosided CA with global rule $\sigma^{-k}\tilde{f}$ is topologically conjugate to the onesides p^{n-1}-shift.

<u>Proof.</u> Assume we are given a vertical sequence of n-1 columns. As in the proof of the theorem we can reconstruct the upper right part of the graph because $k<n$. But using $k>0$ and the fact that f is left permuting we can reconstruct the upper left part in a symmetric way. ∎

<u>Corollary 5.2.</u> Let $f,g \in F(n)$ be local rules which are both right permuting and $\tilde{f}$ and $\tilde{g}$ the corresponding global rules for onesided CA. Then $\tilde{f}$ and $\tilde{g}$ are topologically conjugate. Similarily the global rules $\sigma^{-k}\tilde{f}$ and $\sigma^{-k'}\tilde{g}$ are topologically conjugate if f and g are both left and right permuting and $0<k,k'<n-1$.

<u>Proof.</u> Both global rules are topologically conjugate to the onesided p^{n-1}-shift. ∎

Next we want to list the sets of conjugate CA for p=2 in F(3). For the description we use algebraic expressions modulo 2. We write $\bar{x}$ for 1-x. For symmetry reasons we only list the rules f with $f(0,0,0)=0$.
The following six transition rules are the only rules in F(3) satisfying the above condition.

$x_1+\bar{x}_1x_2+x_3$ (rule 86)

x_1+x_3 (rule 90)

$x_1x_2+x_3$ (rule 106)

$x_1+x_2+x_3$ (rule 150)

$x_1\bar{x}_2+x_3$ (rule 154)

$\bar{x}_1x_2+x_3$ (rule 166)

If we combine the two examples given above we can explicitely see the transformation mapping the first line of the first example into the fist line of the second one.

Note that only the rules 90 and 150 correspond to additive CA, see, for instance, [4]. Hence, additivity is not an invariant under topological conjugacy.

Also only the rules 90 and 150 are left permuting, hence there are only two twosided CA in F(3) conjugate to the 2^2-shift.

4. Application to periodic points

In general it is a nontrivial problem to determine all configurations which are periodic with a certain period under a given CA rule. But if the considered CA is topologically conjugate to a shift then periodic configurations correspond to points periodic under the shift and for the shift on p symbols there are just p^n points of period n.

Next we give two examples.

Example 7. Consider again the local rule 90, i.e. $f(x_1,x_2,x_3)=x_1+x_3$. Let $x=(1,0,1,0,1,0,\ldots)$. Then $\tilde{f}^k(x)=(0,0,0,\ldots)$ for $k=1,2,\ldots$ Hence, x is not a periodic point, but eventually periodic.

Example 8. Consider the following local rule $f\in F(3)$:

$f(x_1,x_2,x_3) = x_1+x_2x_3$. For all satisfying $x_{2k}=0$ for all $k\in\mathbb{N}$ we have $\tilde{f}(x)=x$. Hence, there are infinitely many fixed points under $\tilde{f}$ which are not eventually periodic as a sequence of co-ordinates.

In general we have:

Proposition 9. Let f be an arbitrary local rule and x a point with periodic sequence of co-ordinates. Then x is eventually periodic under f.

Proof. Let q be the minimal period of the sequence of co-ordinates of x. Then f(x) again has period q (here q has no more to be minimal). There are only finitely many blocks of length q, hence, x has to be eventually periodic.

■

The following proposition gives a partial converse.

Proposition 10. Let $f \in F(n)$, $n>1$, be right permuting and x a point periodic under f. Then the sequence of co-ordinates of x is eventually periodic.

Proof. For simplicity we assume that x is a fixed point. The same arguments work in the general case if we consider a suitable power of $\tilde{f}$.

If x is fixed and f right permuting, then $x_1 = f(x_1, \ldots, x_n)$ and x_n is a function of $x_1, \ldots, x_{n-1}$. Now there are only finitely many blocks of length n-1, hence, the co-ordinates of x must be eventually periodic.

■

5. Vertical subshifts

In theorem 5 and the following applicatios we always used the permuting properties of the code. In this section we will show that some of the results are true for some non-permuting CA if we replace the vertical shift by a certain subshift.

Let us recall the definition of a subshift of finite type [2]. Consider a set of words W (of finite length) of symbols of S. Define a subset X of $S^{\mathbf{N}}$ by the restriction that the sequence of co-ordinates of a point of X does not contain any finite subsequence which is a word of W. This subset $X \subset S$ is closed and invariant under the shift. If the set W is finite the shift restrited to X is called a subshift of finite type.

Let us consider the following special rule $f \in F(4)$ for $p=2$:

$$f(x_1,x_2,x_3,x_4) = x_2+\bar{x}_1 x_3 \bar{x}_4 \pmod 2$$

(again we let $\bar{x}=1-x$).

Let σ be the left shift on the sequence of co-ordinates. Then the rule $\tilde{f}$ corresponding to the above f satisfies $\tilde{f}^2=\sigma^2$ [1]. If Φ_2 is defined as above this relation can be used to construct Φ_2^{-1} as follows. We have

$$x_k^1 = x_{k-2}^{1+2} \text{ for } k\geq 2, \text{ hence},$$

$$x_{2k}^0 = x_0^{2k} \quad \text{and} \quad x_{2k+1}^0 = x_1^{2k}.$$

Using these equations and given a vertical sequence of pairs we can reconstruct the uper right part as before. But now we have to check that the constructed object is indeed the graph of our CA for a given initial configuration. Therefore we have to check that arbitrary five elements positioned as

$$\begin{matrix} v & w & x & y \\ z & & & \end{matrix}$$

satisfy $z=f(v,w,x,y)$. By the above construction it is enough to check that for v in the first and second column. Equivalently any seven numbers positioned as

$$\begin{matrix} t & u & v & w & x \\ y & z & & & \end{matrix}$$

have to satisfy $y = f(t,u,v,w)$

and $z = f(u,v,w,x)$. (3)

These elements appear in the first two columns as follows:

$$\begin{matrix} t & u \\ y & z \\ v & w \\ * & * \\ x & * \end{matrix}$$

(here * denote arbitrary elements).

Hence, if a vertical sequence is given as the image of a certain initial state under $\tilde{f}$, the equations (3) give restrictions on the set of words of length 5 in this vertical sequence. On the other hand, if these restrictions are fulfilled, our reconstruction necessarily gives the graph of f for some initial configuration. Hence, $\tilde{f}$ is conjugate to a subshift of finite type.

We consider the two cases $f(w,x,y,z)=\bar{x}$ and $f(w,x,y,z)=x$. In the first case, we have $w\bar{y}\bar{z}=1$, i.e. w=0, y=1 and z=0. In the second case we have $\bar{w}\bar{y}z=0$, i.e. one of the factors has to be zero. Hence, the following combinations are impossible:

$$\begin{matrix} 1 & x & * & * \\ \bar{x} & & & \end{matrix} \qquad \begin{matrix} * & x & 0 & * \\ \bar{x} & & & \end{matrix} \qquad \begin{matrix} * & x & * & 1 \\ \bar{x} & & & \end{matrix} \qquad \begin{matrix} 1 & x & 0 & 1 \\ x & & & \end{matrix}$$

Considering both equations of (3) we get the following set of forbidden words (* denotes arbitrary elements, a is a element fixed in a block):

$$\begin{matrix} 1 & a \\ \bar{a} & * \end{matrix} \qquad \begin{matrix} * & a \\ \bar{a} & * \\ 0 & * \end{matrix} \qquad \begin{matrix} * & a \\ \bar{a} & * \\ * & 1 \end{matrix} \qquad \begin{matrix} 0 & a \\ a & * \\ 1 & 0 \end{matrix} \qquad \begin{matrix} * & 1 \\ * & a \\ \bar{a} & * \end{matrix} \qquad \begin{matrix} * & a \\ \bar{a} & 0 \end{matrix} \qquad \begin{matrix} * & a \\ \bar{a} & * \\ * & * \\ 1 & * \end{matrix} \qquad \begin{matrix} * & 0 \\ * & a \\ a & 1 \\ 0 & * \end{matrix}$$

<u>References.</u>

[1] Amaroso, S; Y.N. Patt: Decision Procedures for Surjectivity and Injectivity of Parallel Maps for Tessellation Structures, J. Comp. Syst. Sci. <u>6</u> (1972) 448-464

[2] Denker, M.; C. Grillenberger and K. Sigmund: Ergodic Theory on Compact spaces, Springer Lecture Notes in Mathematics 527

[3] Hedlund, G.A.: Endomorphisms and automorphisms of the shift dynamical system, Math. Systems Th. <u>3</u> (1969)

[4] Wolfram, S.: Statistical Mechanics of Cellular Automata, Rev. Mod. Phys. <u>55</u> (1983) S. 601-644

ON A GEOMETRIC CHARACTERIZATION OF DIFFERENTIAL-ALGEBRAIC EQUATIONS

Sebastian Reich [1)]

1 Introduction

Many physical systems are described by differential-algebraic equations (abbreviated DAEs) of the form:

$$x' = f(x,y) \tag{1a}$$

$$0 = g(x,y) \tag{1b}$$

where $x \in R^k$, $y \in R^m$, $f: R^k x R^m \to R^k$, $g: R^k x R^m \to R^m$. For example, it is well-known that electronic circuits may be modeled using equations of the form (1). Indeed, for electronic circuits, (1a) may always be written $q' = i_C$ and $\phi' = u_L$ where q, i_C denote capacitor charges and currents while ϕ, v_L denote inductor fluxes and voltages. The Equations (1b) then represent the Kirchhoff constraints together with the element constitutive relations.

For a long time, it has been realized that the analysis of DAEs of the form (1) is far from being trivial [4,8,10]. This is mainly due to the fact that the solutions of (1) are restricted to a lower dimensional manifold which, in general,

(i) is a submanifold of the set

$$M := \{(x,y):\ 0 = g(x,y)\},$$

and

(ii) cannot be viewed as a linear space by globally coordinatising it.

To overcome this bottleneck, several concepts for the analysis of DAEs have been developed in the past [3,7,13]. Among these the geometric approach, based on the theory of vector fields, seems to us the most promising. Therefore, by imploying ideas of Rheinboldt [13], we regard DAEs as an implicit description of vector fields on manifolds, and have introduced in [11,12] the notion of a regular DAE as a consequence of this approach.

This paper contains a summary of the most important definitions and results concerning regular DAEs. In order to illustrate our concepts, we discuss very detailed in Chapter 4 the extremely simple, however appealing circuit example of Fig. 1.

---

1) Technische Universität Dresden, DDR-8027, Mommsenstr. 13

2 Notions

In the sequel, we restrict our study to quasilinear, time-invariant DAEs of the type

$$A(z)z' = f(z) \qquad (2)$$

where $z \in R^n$, $f: R^n \to R^n$, $A: R^n \to L(R^n)$. This restriction simplifies the results significantly and is motivated by the fact that one can easily generalize the results obtained in this paper to the time-varying case. This can be done by the well-known techniques for reducing a time-varying DAE into a time invariant one [13]. Furthermore, with $z = (x,y)$, (1) is a special case of (2).

By a solution of a DAE (2) we mean a differentiable mapping $c: I \to R^n$ (where I denotes an intervall in the real line) such that for all $t \in I$

$$A(c(t))dc/dt(t) = f(c(t)) \quad .$$

Throughout this paper, we use the following notions. If M is a differentiable manifold then TM denotes the tangent bundle of M and T_xM the tangent space of M at $x \in M$. A vector field on a differentiable manifold M is a mapping $v: M \to TM$ such that $v(x) \in T_xM$ for all $x \in M$. By a solution of a vector field $v: M \to TM$ we mean a differentiable mapping $w: I \to M$ such that for all $t \in I$

$$dw/dt(t) = v(w(t)) \quad .$$

Let $f: M \to R^n$ be a differentiable mapping from a manifold M into R^n. Then we denote the derivative of the mapping f by $Df: TM \to R^n$ and the derivative at $x \in M$ by $Df(x) \in L(T_xM, R^n)$. Additionally, if M is a submanifold of R^n, then let $j: M \to R^n$ be the natural injection. (For an introduction in the theory of manifolds and vector fields we refer to [1].)

Because we like to give a geometric interpretation of DAEs, the set defined in the next definition will turn out to be of special importance.

DEFINITION 1. Let a DAE of the type (2) be given. Then we call the set

$$N := \{ (z,z'): A(z)z' = f(z) \}$$

the corresponding set of the DAE.

Note, in Definition 1, the DAE (2) is considered as a non-linear system of equations in the variables z and z'. For that reason, the corresponding set N of the DAE is a subset of $R^n x R^n$. Clearly, a differentiable mapping $c: I \to R^n$ is a solution of this DAE if and only if $(c(t), dc/dt(t)) \in N$ for all $t \in I$. Therefore, with respect to the solutions of the DAE, we can instead consider the corresponding set. This fact will be explored further in Chapter 3.

In [13], Rheinboldt developed the idea of considering DAEs as an implicit description of vector fields on manifolds. Inspired by this, we have proposed in [11,12] the following definition of a regular DAE.

DEFINITION 2. Let a DAE of the type (2) be given. Then this DAE is called a regular DAE if there is a differentiable submanifold M of R^n and a vector field v: M->TM such that a differentiable mapping w: I->M is a solution of the vector field v if and only if the mapping $c := j \circ w: I \to R^n$ is a solution of the DAE.

The manifold M is called then the configuration space and the vector field v the corresponding vector field of the DAE. ▪

3 Constraint Manifolds and Regular DAEs

In this section, we will derive sufficient conditions for the regularity of a DAE. Furthermore, we state a technique by means of which, one can obtain for a given regular DAE the configuration space M and the corresponding vector field v. To start with, we make the following observation.

OBSERVATION. Let N be the corresponding set of a given DAE (2). We consider now the projection of this set onto the first component, that is the set

$$M_1 := p_1(N) \ .$$

The set M_1 is a subset of R^n and reflects the algebraic constraints on the solutions of the DAE. Let us assume that M_1 is a differentiable manifold. Clearly, under this assumption, a differentiable mapping $c: I \to R^n$ is a solution of the given DAE if and only if

$$(c(t), dc/dt(t)) \in (N \cap TM_1)$$

for all $t \in I$. In general, $(N \cap TM_1)$ will be a subset of N. Therefore, we consider the set

$$M_2 := p_1(N \cap TM_1) \ .$$

If now the set M_2 is a differentiable manifold as well, we conclude again that a differentiable mapping $c: I \to R^n$ is a solution of the DAE if and only if

$$(c(t), dc/dt(t)) \in (N \cap TM_2)$$

for all $t \in I$. This process can be continued as long as the set

$$M_i := p_1(N \cap TM_{i-1})$$

is a differentiable manifold and may be stopped if $M_i \equiv M_{i-1}$.

This observation leads us in a natural way to the following definition.

DEFINITION 3. Let N be the corresponding set of a given DAE (2). We define a family $(M_i)_{i=0,...,s}$ of manifolds M_i by the recursion:

(i) $M_0 := R^n$

(ii) $M_{i+1} := P_1(N \cap TM_i)$ $(i=1,...,s-1)$

where s is the largest integer such that the sets M_i are differentiable manifolds and $M_{s-1} \equiv M_s$. In case $M_1 \equiv R^n$, we define $s = 0$.

We call the family (M_i) the family of constraint manifolds and the integer s the degree of the given DAE.

REMARKS. In [14], Szatkowski proposed a similar definition independently of us. In [13], Rheinboldt described a procedure for obtaining a family of manifolds as the solutions of an overdetermined system of equations. One can show that this family is identical with our family of constraint manifolds. Therefore, our definition can be considered as a geometric interpretation of the procedure given there. In [12], we have proved that $s \leq n$ for any DAE of the type (2). ▪

By means of the family of constraint manifolds, we can state sufficient conditions for the regularity of a DAE.

THEOREM. Let N be the corresponding set, (M_i) the family of constraint manifolds, and s the degree of a given DAE (2). Then this DAE is regular if the condition

C: $(N \cap T_xM_s)$ contains exactly one element for each $x \in M_s$

is satisfied.

Under this condition, the configuration space M of the DAE is given by $M \equiv M_s$, and the corresponding vector field $v: M \to TM$ is for all $x \in M$ defined by

$$\{v(x)\} = (N \cap T_xM) .$$

PROOF. The Condition C insures that the mapping $v: M \to TM$, defined by

$$\{v(x)\} = (N \cap T_xM)$$

for all $x \in M$, is a vector field on M. Furthermore, from our observation, it is readily derived that a differentiable mapping $w: I \to M$ is a solution of the vector field v if and only if the mapping $c := j \circ w: I \to R^n$ is a solution of the given DAE.

Clearly, the above theorem allows for the development of a solution theory for regular DAEs by 'translating' related results of vector fields to regular DAEs. To do this is beyond the scope of this paper, and we refer instead to [11].

In several papers [3,6,9], the properties of DAEs are characterized by an integer called the index. We want to state here a definition of the index of a DAE as well. In [11,12], it is shown that our definition is in essence equivalent to the one given in [3,6,9].

DEFINITION 4. Let N be the corresponding set, (M_i) the family of constraint manifolds, and s the degree of a DAE (2). Then this DAE is called a DAE of index s if it satisfies the Condition C of the above theorem.

REMARK. We like to emphasize that any DAE (2), for which an index can be defined, is a regular DAE according to the above theorem.

4 A Simple RC-Circuit

Throughout this chapter, let us consider the RC-circuit of Fig. 1 where we assume the voltage-current relation S of the non-linear resistor to be given by

$$S := \{(u_R, i_R): 0 = g(u_R, i_R)\}$$

with $g: R\times R \to R$ and $Dg(u_R, i_R)$ has full rank for all $(u_R, i_R) \in R\times R$. This implies that the set S is a differentiable manifold and that the time behavior of the RC-circuit can be modeled by the DAE

$$q' = i \qquad (3a)$$

$$0 = g(q, -i) \qquad (3b)$$

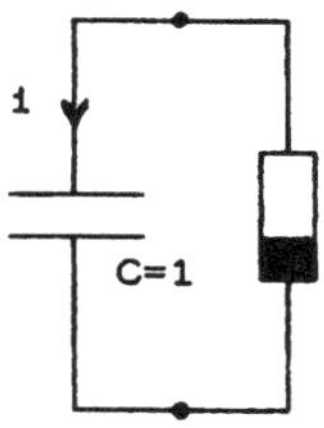

Fig. 1

Obviously, in general, Equation (3b) cannot be solved globally for the variable i. Consequently, the state equations [4] does not exist in any case. Therefore, we apply now the more general theory of regular DAEs to the DAE (3).

Clearly, by definition, the constraint manifold M_1 is given by

$$M_1 = \{(q, i): 0 = g(q, -i)\}.$$

Now, a point $(q, i) \in M_1$ is an element of the set M_2 if and only if

$$(N_{(q,i)} \cap T_{(q,i)}M_1) \neq \emptyset \qquad (4)$$

where

$$N_{(q,i)} = \{(q', i'): q' = i\},$$

$$T_{(q,i)}M_1 = \{(q', i'):\ 0 = Dg(q,-i)(i', q')T\} \quad .$$

It is readily shown that we have

$$(N_{(q,i)} \cap T_{(q,i)}M_1) = \emptyset$$

if and only if $D_2g(q,-i) = 0$ and $i \neq 0$. This is illustrated in Fig. 2a. In the sequel, we call a point $(q,i) \in M_1$ with these properties a singular point.

Obviously, in case (4) is satisfied, then we have either

a) $D_2g(q,-i) = 0$ and $i = 0$ (Fig. 2b)

or

b) $D_2g(q,-i) \neq 0$ (Fig. 2c) .

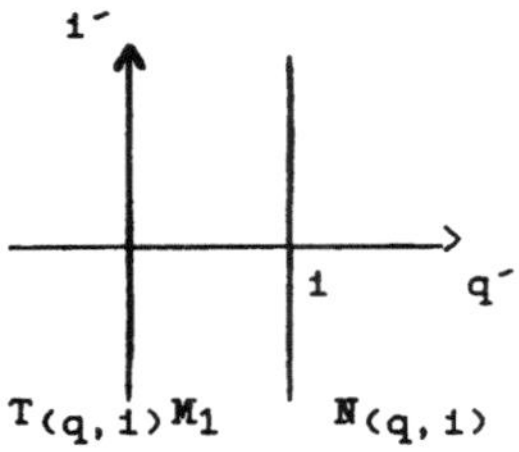

Fig. 2a

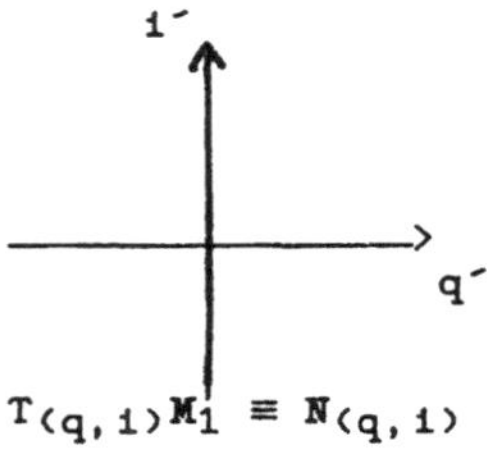

Fig. 2b

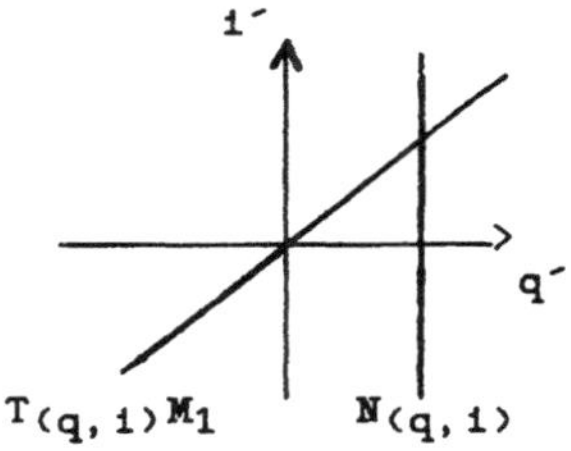

Fig. 2c

For b), we obtain that the set

$$N_{(q,i)} \cap T_{(q,i)}M_1$$

contains exactly one element, while for a), this set is diffeomorphic to the real line $\mathbb{R}$. Therefore, we call a point $(q,i) \in M_1$ regular if b) holds, and degenerated if a) holds.

According to the remarks in Chapter 3, we conclude that the degree s of the DAE (3) cannot exceed $s = 2$.

Now we discuss the results obtained so far by means of the specific voltage-current relations given in the Fig. 3a-e.

EXAMPLE 1. First we consider the RC-circuit of Fig. 1 with the voltage-current relation of the non-linear resistor depicted in Fig. 3a. Clearly, we have

$$M_2 \equiv M_1,$$

that is, every point $(q,i) \in M_1$ is a regular point. Consequently, the DAE (3) is regular and of index 1. ▪

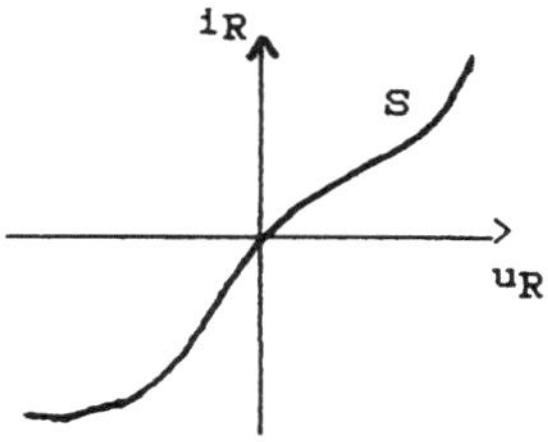

Fig. 3a

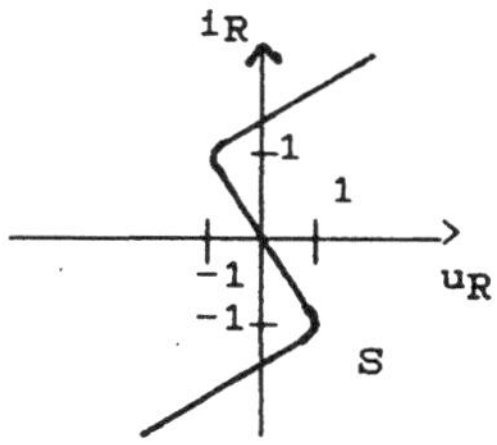

Fig. 3b

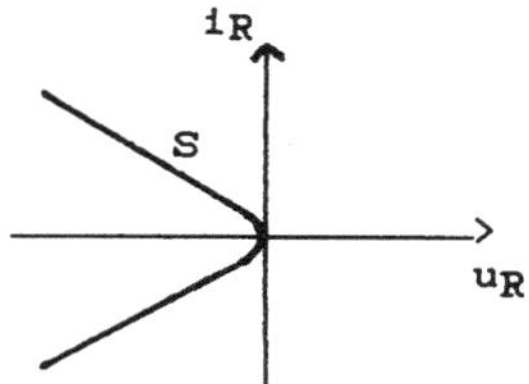

Fig. 3c

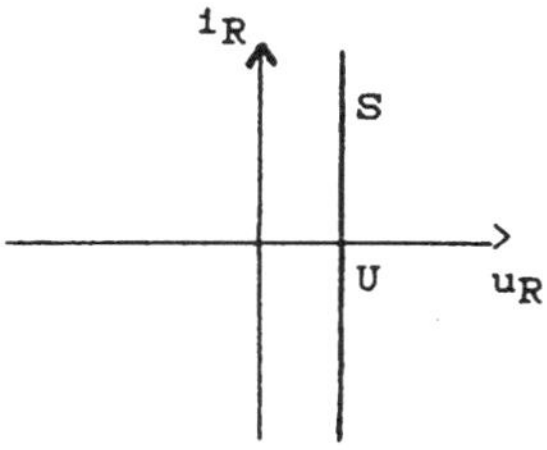

Fig. 3d

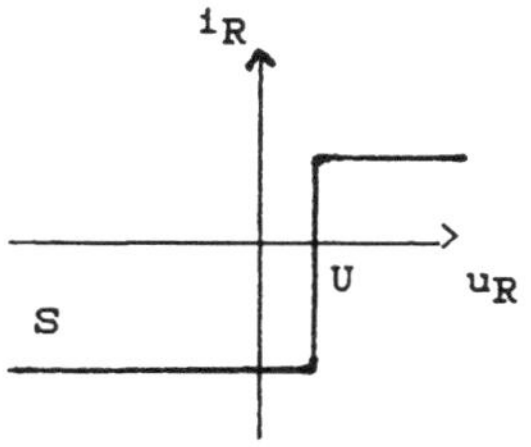

Fig. 3e

EXAMPLE 2. Now, we consider the voltage-current relation S of Fig. 3b. We obtain

$$M_2 = M_1 \setminus \{(-1,-1),(1,1)\}$$

that is, (-1,-1), (1,1) are singular points and all the other points belonging to M_1 are regular. Consequently, the DAE (3) is in this case regular and of index 2. The vector field on M_2 is depicted schematically in Fig. 4a. Obviously, no solution can be extented beyond the two points (-1,-1) and (1,1). Therefore, such points are called impasse points [4,5]. Impasse points are an important phenomena found in many papers (see, for example, [5]) in connection with the so called jump behavior. ∎

EXAMPLE 3. Let the voltage current relation S be given by Fig. (3c). Similar to Example 1, we obtain

$$M_2 \equiv M_1 \quad .$$

However, the point (0,0) is a degenerated point, and as a consequence of this, the Condition C in Chapter 3 does not hold because

$$N_{(0,0)} \cap T_{(0,0)}M_2$$

is diffeomorphic to the real line R.

EXAMPLE 4. Now we consider the voltage-current relation S of Fig. 3d. The resulting circuit is equivalent to the circuit which we obtain by replacing the non-linear resistor by a dc voltage-source $u(t) = U$. With exception of the point $(U,0)$ all other points belonging to M_1 are singular, and we get

$$M_2 = \{(U,0)\}.$$

The point $(U,0)$ is degenerated. But, because $\dim(M_2) = 0$, we have

$$(N_{(U,0)} \cap T_{(U,0)}M_2) = \{(0,0)\} .$$

Accordingly, the Condition C is satified, and the DAE (3) is regular with the index 2 (see Fig. 4b).

EXAMPLE 5. Finally, consider the voltage-current relation S given in Fig. 3e. The circuit specified by this characteristic is closely related to systems with variable structure and sliding mode [2]. So far such systems have been modeled by differential inclusions [2]. By means of this example, we argue that those systems can be considered as higher index DAEs as well.

The constraint manifold M_2 is depicted in Fig. 4c. Again the point $(U,0)$ is a degenerated point. However the same arguments like in Example 4 yield that the DAE (3) is regular and of index 2. We like to remark that the sliding mode $c: I \to R \times R$ is in this case simply given by

$$c(t) = (U,0) \quad .$$

∎

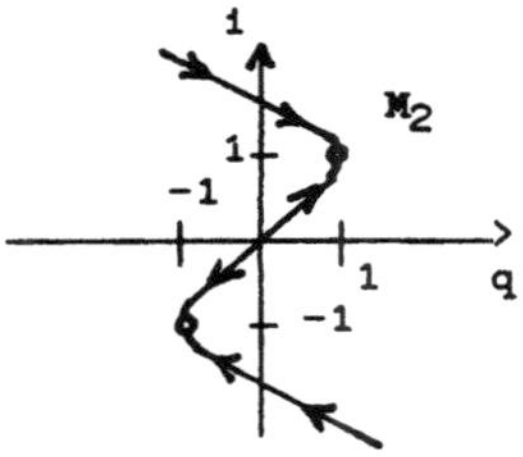

Fig. 4a

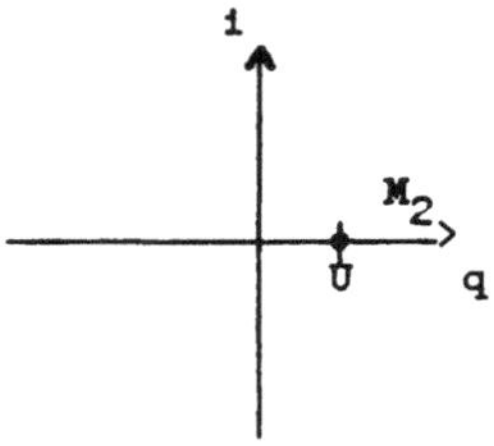

Fig. 4b

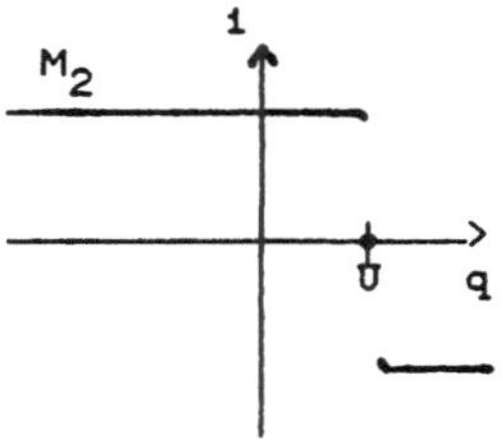

Fig. 4c

5 Concluding Remarks

By means of a simple RC-circuit, we have shown that the properties of DAEs can be very non-trivial. Therefore, in this and other papers, we have stated a method for a characterization of DAEs in purely geometric terms. However, many open questions still remain in connection with this method. For example, the so far obtained results should be generalized to non-autonomous DAEs, and necessary conditions for the differentiability of the constraint manifolds should be derived in terms of transversality [8].

REFERNCES

[1] Arnol'd, V.I.: Ordinary differential equations. - Cambridge, Mass.: MIT Press, 1981

[2] Aubin, J.P.; Cellina, A.: Differential inclusions. - Berlin, Heidelberg, New York, Tokoy: Springer, 1984

[3] Brenan, K.E.; Campbell, S.L.; Petzold, L.R.: The numerical solution of initial value problems in ordinary differential- algebraic equations. - North Holland Publishing Co., 1989

[4] Chua, L.O.: Introduction to non-linear network theory. - New York: McGraw Hill Book Company, 1969

[5] Chua, L.O.; Deng, A.: Impasse points. - in: Int. J. of Circuit Theory and Appl., Vol. 17(1989), pp. 213-235

[6] Gear, C.W.: DAEs index transformations. - in: SIAM J. Sci. Stat. Comp., Vol. 9(1988)1, pp. 39-47

[7] Griepentrog, E.; März, R.: DAEs and their numerical treatment. - Leipzig: Teubner Verlagsgesellschaft, 1986

[8] Haggman, B.C.; Bryant, P.R.: Solutions of singular constrained differential equations. - in: IEEE Trans. CAS, Vol. 31(1984), pp. 1015-1029

[9] Lötstedt, P.; Petzold, L.R.: Numerical solution of nonlinear differential equations with algebraic constraints I. - in: Math. Comp., Vol. 46(1986)174, pp. 491-516

[10] Petzold, L.R.: Differential/algebraic equations are not ODEs. - in: SIAM J. Sci. Stat. Comp., Vol. 3(1982), pp. 367-384

[11] Reich, S.: Beitrag zur Theorie der Algebrodifferentialgleichungen. - TU Dresden: thesis (manuscript), 1989

[12] Reich, S.: On a geometrical interpretation of DAEs. - to appear in: Circuits, Systems, and Signal Processing Journal

[13] Rheinboldt, W.C.: Differential-algebraic systems as differential equations on manifolds. - in: Math. Comp., Vol. 43(1984)168, pp. 1473-1482

[14] Szatkowski, A.: On generalized dynamical systems. - in: Proc. ECCTD-89, Brighton, 1989, pp. 803-808

ASYMPTOTIC BEHAVIOUR OF TRAJECTORIES OF DYNAMICAL SYSTEMS ON THE CYLINDER

V. Reitmann [1)]

1. Introduction. Continuous-time or discrete-time dynamical systems on the cylinder are also called phase systems ([3, 5, 8 - 10, 12, 17]). Such systems describe models of phase synchronization ([2, 17]), coupled Josephson junctions ([3, 14, 16]), and other systems with angular coordinates ([1, 3, 5]). Phase systems represent special systems with multiple equilibria. A central issue of investigating such systems is to provide sufficient conditions for the solutions to converge. In [5] it is shown what advantages are to be gained if we consider such systems as control systems. This way, for phase systems we are able to obtain results concerning dissipativity, boundedness and convergence with the use of positively invariant quadratic cones ([5, 10, 13]).

A new approach to the problem of complete stability, the nonlinear reduction method, was developed by Leonov ([5, 11, 13]) who expressed stability criteria for the n-dimensional system in terms of the frequency domain response of the linear part and in terms of solutions of second-order comparison systems.

In [18] convergence results are shown for differential equations with multiple equilibria using the eigenvalues of the symmetrized Jacobian matrix of the differential equation and positively invariant compact sets. For certain classes of phase systems with a gradient nonlinearity on a noncompact manifold the convergence of a trajectory is equivalent to the boundedness ([1, 16]). It was already in the paper [2] that Bakaev and Guž have shown the possibility of proving the convergence of trajectories of dynamical systems with periodic in the state nonlinearities by means of periodic Ljapunov functions. This kind of applying Ljapunov functions is known as Bakaev-Guž technique. Using the Jakubovič-Kalman lemma ([5, 6]) for the solvability of a special matrix inequality, convergence criteria were proved for phase systems in the form of feedback equations.

Questions concerning convergence of solutions of phase systems can be well formulated in terms of factor-spaces. In his paper [4] Brockett proposes a method of investigating the convergence of differential equations on a flat Riemannian manifold. This method is based on assumptions on the periods of closed one-forms and on the fundamental group of the manifold for the existence of Ljapunov functions defined on the ω-limit set of semicompact trajectories.

---

1) Dresden University of Technology, Department of Mathematics, Mommsenstr. 13, Dresden, GDR - 8027

Realizing this abstract approach for a class of control phase systems in the second canonical form ([5]) the author [4] provides sufficient frequency domain conditions for the trajectories to converge early deduced in [5].

Our result requires exact one-forms only, the proof dispenses with the invariance principle of La Salle. We also obtain an analogue result for discrete-time systems on the cylinder. The principal tool necessary for the construction of exact one-forms in this paper is the Jakubovič-Kalman lemma ([5, 6]).

Let us consider a differential equation of the form

$$\dot{x} = f(x) \tag{1}$$

where $f : \mathbb{R}^n \to \mathbb{R}^n$ is a continuous function. Let associate with $\mathbb{R}^n$ the standard inner product. Suppose that all solutions of (1) exist on $[0,+\infty)$. We denote by G a set of vectors $d \in \mathbb{R}^n$ ("periods") with $f(x + d) = f(x)$ for all $x \in \mathbb{R}^n$. If $G \neq \{0\}$, then such a system is called a continuous-time phase system. Suppose that the dimension of the linear space generated by G is less than n. On these assumptions the equivalence relation $x \sim y \iff x = y + d$ for some $d \in G$ gives the structure of a Riemannian manifold $\mathbb{R}^n/G$ (with the flat metric from $\mathbb{R}^n$) which we call a flat cylinder ([7, p. 199]). Thus the differential equation (1) can also be looked upon as defining a vector field on the tangent bundle on $\mathbb{R}^n/G$.

In addition to the differential equation (1) we consider the discrete-time system

$$x(k + 1) = F(x(k)), \qquad k = 0,1,\ldots \tag{2}$$

with a continuous function $F : \mathbb{R}^n \to \mathbb{R}^n$.

Let us suppose, that there exist vectors $d \in \mathbb{R}^n$, $d \neq 0$, such that $F(x + d) = F(x) + d$ for all $x \in \mathbb{R}^n$. We call system (2) with such a property a discrete-time phase system.

Defining the group G as above we can look at the system (2) as a system on the flat cylinder $\mathbb{R}^n/G$.

2. Dynamical systems on metric space

In generalization of the situation treated in the foregoing section we interpret system (1) as a vector field on a Riemannian manifold (M,g). For each $a \in M$ we denote by $t \to x(t,a)$ a solution of (1) for $t \geq 0$ satisfying $x(0,a) = a$. The set $\{x(t,a),\ t \geq 0\}$ is called a semi-trajectory of (1). A point $a \in M$ with $x(t,a) = a$ for $t \geq 0$ is called an equilibrium point of (1). The semi-trajectory of a solution x is said to converge to an equilibrium a if $x(t) \to a$ for $t \to \infty$. A semi-trajectory is said to be bounded if its closure in M is compact. If ω_1 and

ω_j are one-forms on M then $\langle \omega_i, \omega_j \rangle$ denotes their inner product.

<u>Theorem 1.</u> Let f be a vector field on M and let ω be the one-form dual to f. Suppose that $\omega = \omega_1 + \omega_2$ is a decomposition of ω such that ω_1 is exact and the following conditions hold:

(i) $\langle \omega_1, \omega \rangle \leq - \alpha$ on M for a continuous function $\alpha : M \to \mathbb{R}_+$.

(ii) If $\int_0^\infty \alpha(x(t))dt < \infty$ for a solution $x(\cdot)$ of $\dot{x} = f(x)$ with bounded semi-trajectory then the semi-trajectory of $x(\cdot)$ converges to an equilibrium.

Then any bounded semi-trajectory of $\dot{x} = f(x)$ converges to an equilibrium.

<u>Proof.</u> Suppose x is a solution with bounded semi-trajectory. Since by assumption the one-form ω_1 is exact we can represent it as the exterior differential $\omega_1 = d\Phi$. We define for $t \geq 0$ the function

$$V(t) = \Phi(x(t)) - \Phi(x(0)).$$

From (i) it follows that

$$V(t) = \int_0^t \langle \omega_1, \omega \rangle dt \leq - \int_0^t \alpha(x(s))ds.$$

By assumption $x(\cdot)$ has a bounded semi-trajectory and Φ is also bounded. Thus we have

$$0 \leqq \int_0^t \alpha(x(s))ds \leqq - V(t)$$

which immediately implies

$$\int_0^\infty \alpha(x(s))ds < \infty.$$

This establishes Theorem 1.

By analogy with Theorem 1 we obtain a corresponding convergence result for the discrete System (2). We suppose in the following that the map $F : M \to M$ is defined on a metric space which is not necessarily a Riemannian manifold. The terms equilibrium, bounded semi-trajectory, and convergence with respect to the discrete system (2) on M correspond to the analogue for the continuous-time system (1) providing that in the above definition the values $t \in \mathbb{R}_+$ are replaced by $\mathbb{Z}_+$.

<u>Theorem 2.</u> Let for the continuous functions $F : M \to M$, $V : M \to \mathbb{R}$ and $\alpha : M \to \mathbb{R}_+$ the following conditions hold:

(i) $V \circ F - V \leqq - \alpha$ on M.

(ii) If $\sum_{k=0}^\infty \alpha(x(k)) < \infty$ for some solution x of $x(k + 1) = F(x(k))$ with bounded semi-trajectory then this semi-trajectory converges to an equilibrium.

Then any bounded semi-trajectory of $x(k + 1) = F(x(k))$ converges to an equilibrium.

<u>Proof.</u> Let x be a solution of $x(k + 1) = F(x(t))$ with bounded semi-trajectory. For any $n \in \mathbb{Z}_+$ it follows from (i) that

$$V(x(n)) - V(x(0)) \leq - \sum_{k=0}^{n-1} \alpha(x(k)).$$

We conclude from this that

$$\sum_{k=0}^{\infty} \alpha(x(k)) < \infty$$

and thus the semi-trajectory converges to an equilibrium. The proof is complete.

3. Continuous time feedback systems on the cylinder

In this section we apply Theorem 1 to a special class of control phase systems (1). For this purpose we investigate continuous-time phase systems in the second canonical form ([4, 5, 9]) as a single-loop continuous feedback system

$$\begin{aligned} \dot{x} &= Ax + b\varphi(\sigma), \\ \dot{\sigma} &= c^*x + \varrho\varphi(\sigma). \end{aligned} \tag{3}$$

Such a system can exhibit every type of dynamics that occurs in ordinary differential equations in $\mathbb{R}^m$.

In this system let A be an $n \times n$-Hurwitzian matrix (i.e. the eigenvalues of A lie in the left half-plane), b and c n-column vectors, ϱ a number ($\varrho \neq c^*A^{-1}b$). We suppose that $\varphi : \mathbb{R} \to \mathbb{R}$ is a continuous Δ-periodic function which possesses a finite number of zeros on $[0,\Delta)$. For a matrix B by B^* is denoted the transpose. Suppose that the pair (A,b) is controllable (i.e. $\det[b,Ab,\dots,A^{n-1}b] \neq 0$). Let $K(p) = c^*(A - pI_n)^{-1}b - \varrho$ denote the transfer function of the linear part associated with (3) defined for complex p with $\det(A - pI_n) \neq 0$ (I_n is the $n \times n$-unit matrix).

Our aim is to show for the phase system (3) that we get the hypotheses of the classical Bakaev-Guž technique ([5]) by using Theorem 1. In contrast to the proof in [4] we use exact one-forms only.

Under the above propositions system (3) is invariant with respect to the transformation $\begin{bmatrix} x \\ \sigma \end{bmatrix} \to \begin{bmatrix} x \\ \sigma+\Delta \end{bmatrix}$ such that this system can be regarded as being defined on $\mathbb{R}^{n+1}/G \cong \mathbb{R}^n \times S^1$ with $G = \{jd,\ j \in \mathbb{Z}\}$ and $d = \begin{bmatrix} 0 \\ \Delta \end{bmatrix}$.

The metric is induced by the metric $dx_1^2 + \ldots + dx_n^2 + d\sigma^2$, where $(x, \sigma \pmod{\Delta})$ are the coordinates in $\mathbb{R}^n \times S^1$. Because of $\varrho \neq c^*A^{-1}b$ the set of equilibria $\{x = 0,\ \sigma = \bar{\sigma} : \varphi(\bar{\sigma}) = 0,\ \bar{\sigma} \in [0,\Delta)\}$ of (3) consists of isolated points only.

Theorem 3. For some parameters ϑ, $\varepsilon > 0$ and $\delta > 0$ let the following inequalities hold:

(a) $\vartheta \operatorname{Re} K(i\theta) - \varepsilon|K(i\theta)|^2 - \delta \geqq 0$ for all $\theta \in \mathbb{R}$.

(b) $4\delta\varepsilon > (\vartheta\nu)^2$ with $\nu = \int_0^\Delta \varphi(\sigma)d\sigma / \int_0^\Delta |\varphi(\sigma)|d\sigma$.

Then any semi-trajectory of system (3) converges to an equilibrium.

Proof. Since by assumption the pair (A,b) is controllable and (a) holds by the Jakubovič-Kalman lemma ([5, 6]) there exists a symmetric $n \times n$-matrix H such that for all $(x,\xi) \in \mathbb{R}^n \times \mathbb{R}$ we have

$$2x^*H(Ax + b\xi) + \vartheta(c^*x + \varrho\xi)\xi + \varepsilon(c^*x + \varrho\xi)^2 + \delta\xi^2 \leqq 0. \tag{4}$$

Using this matrix H we define on $\mathbb{R}^n \times S^1$ equipped with the coordinates $(x, \sigma \pmod{\Delta})$, the one-form

$$\omega_1 = 2x^*Hdx + \vartheta(\varphi - \nu|\varphi|)d\sigma. \tag{5}$$

Because of the choice of ν in (b) ω_1 is exact, and, therefore, the relation $\omega_1 = d\Phi$ holds with

$$\Phi(x,\sigma) = x^*Hx + \vartheta \int_0^\sigma (\varphi(r) - \nu|\varphi(r)|)dr. \tag{6}$$

Immediately following from (4) for all $(x,\sigma) \in \mathbb{R}^n \times S^1$ is the estimate

$$\langle \omega_1, \omega \rangle + \vartheta\nu|\varphi(\sigma)|(c^*x + \varrho\varphi(\sigma)) + \varepsilon(c^*x + \varrho\varphi(\sigma))^2 + \delta\varphi(\sigma)^2 \leqq 0. \tag{7}$$

Because of (b) we can select constants $\delta_0 > 0$ and $\varepsilon_0 > 0$ such that

$$\vartheta\nu|\xi|(c^*x + \varrho\xi) + \varepsilon(c^*x + \varrho\xi)^2 + \delta\xi^2 \geqq \delta_0\xi^2 + \varepsilon_0(c^*x + \varrho\xi)^2$$

for all $(x,\xi) \in \mathbb{R}^n \times \mathbb{R}$. Using this and (7) we conclude for all $(x,\sigma) \in \mathbb{R}^n \times S^1$, that

$$\langle \omega_1, \omega \rangle \leqq -\delta_0\varphi^2(\sigma). \tag{8}$$

Since A is an Hurwitzian matrix and φ is continuous, any solution $x(\cdot)$, $\sigma(\cdot)$ of (3) exists on $[0,+\infty)$ and admits a bounded semi-trajectory on $\mathbb{R}^n \times S^1$. Then by (8) we obtain for such a solution

$$\int_0^{+\infty} \varphi^2(\sigma(t))dt < \infty.$$

It remains to show that $x(t) \to 0$ and $\sigma(t) \to \sigma^*$ for a zero of φ. Because φ is continuous and Δ-periodic and $\dot{\sigma} = c^*x + \varrho\varphi$ is bounded, the function $\varphi(\sigma(\cdot))$ is uniformly continuous on $[0,+\infty)$. Thus from

$$\int_0^\infty \varphi^2(\sigma(t))dt < \infty$$

we have $\varphi(\sigma(t)) \to 0$ for $t \to \infty$.

Since by assumption the zeros of φ are isolated the foregoing relation implies $\sigma(t) \to \sigma^*$ with $\varphi(\sigma^*) = 0$.

Recalling that A is an Hurwitzian matrix it immediately follows from the first equation of (3) that $z(t) \to 0$ for $t \to \infty$. We may apply Theorem 1 to conclude that all semi-trajectories converge.

4. Discrete-time feedback systems on the cylinder

Similarly to the continuous-time case we investigate a discrete control phase system in the second canonical form ([8, 10])

$$\begin{aligned} x(k+1) &= Ax(k) + b\varphi(\sigma(k)), \\ \sigma(k+1) &= \sigma(k) + c^*x(k) + \varrho\varphi(\sigma(k)), \qquad k = 0,1,\dots . \end{aligned} \tag{9}$$

We show to apply Theorem 2 in order to conclude the convergence of all semi-trajectories. In this system let A be an $n \times n$-matrix, with the eigenvalues of which having moduli less than 1, b and c be n-vectors and ϱ be a number ($\varrho \neq c^*(A - I_n)^{-1}b$). Let us assume further that $\varphi : \mathbb{R} \to \mathbb{R}$ is a Δ-periodic function which is Lipschitz on $\mathbb{R}$ with a constant L, and possesses a finite number of zeros on $[0,\Delta)$. Suppose that (A,b) is controllable and K is the function defined in section 3.

Under the conditions described for system (9) we can interpret it also on the flat cylinder $\mathbb{R}^{n+1}/G \cong \mathbb{R}^n \times S^1$, where G is given in section 3.

The first extension of the Bakaev-Guž technique for discrete phase systems of the type (9) was done in [8].

In addition to the results of [8] the following theorem provides new conditions under which any semi-trajectory of (9) converges to an equilibrium. To establish this result on the cylinder $\mathbb{R}^n \times S^1$ by means of Theorem 2 we use an exact one-form of the type (5) in a similar manner as in the proof of Theorem 3 for constructing the function V of Theorem 2. The result [8] mentioned above can also be obtained with the help of another one-form employing Theorem 2. The following theorem can be viewed as the discrete analogue of Theorem 3.

Theorem 4. For the parameters ϑ, $\varepsilon > 0$ and $\delta > 0$ suppose:

(i) $\vartheta \operatorname{Re} K(p) - \varepsilon|K(p)|^2 - \delta \geqq 0$ for all $p \in \mathbb{C}$ with $|p| = 1$.

(ii) $4(\varepsilon - \vartheta L(1 + \nu)\delta) > (\nu\vartheta)^2$.

Let ν be as described above by the relation (b) in Theorem 3 with φ from (9).

Then any semi-trajectory of system (9) converges to an equilibrium.

<u>Proof.</u> It may be argued exactly the same way as in proof of Theorem 3. Since the pair (A,b) is controllable and the inequality (i) holds, all hypotheses of the Jakubovič-Kalman-Szegö lemma [6] are given. Because of this lemma there exists an $n \times n$-matrix $H = H^*$ such that

$$(Ax + b\xi)^* H(Ax + b\xi) - x^*Hx + \vartheta\xi(c^*x + \varrho\xi) + \varepsilon(c^*x + \varrho\xi)^2 + \delta\xi^2 \leqq 0 \qquad (10)$$

for all $(x,\xi) \in \mathbb{R}^n \times \mathbb{R}$. Using this matrix H on $\mathbb{R}^n \times S^1$ we define the exact one-form ω_1 from (5) and get $\omega_1 = dV$ with $V = \Phi$ from (6). We denote by $(x,\sigma) = z$ the coordinate $(x,\sigma) \in \mathbb{R}^n \times S^1$ and by $z(k + 1) = F(z(k))$ the iteration (9). Then by (10) we obtain the inequality

$$V(F(z)) - V(z) - \vartheta \int_{\sigma}^{\sigma+c^*x+\varrho\varphi(\sigma)} (\varphi(r) - \nu|\varphi(r)|)dr$$
$$+ \vartheta\varphi(\sigma)(c^*x + \varrho\varphi(\sigma)) + \varepsilon(c^*x + \varrho\varphi(\sigma))^2 + \delta\varphi(\sigma)^2 \leqq 0 \qquad (11)$$

for all $(x,\sigma) \in \mathbb{R}^n \times S^1$. In applying a mean value theorem to the integral and using the Lipschitz continuity of φ we get for all $(x,\sigma) \in \mathbb{R}^n \times S^1$ the relations

$$\int_{\sigma}^{\sigma+c^*x+\varrho\varphi(\sigma)} (\varphi(r) - \nu|\varphi(r)|)dr = (\varphi(\bar\sigma) - \nu|\varphi(\bar\sigma)|)(c^*x + \varrho\varphi(\sigma))$$
$$= (\varphi(\sigma) - \nu|\varphi(\sigma)|)(c^*x + \varrho\varphi(\sigma))$$
$$+ [(\varphi(\bar\sigma) - \nu|\varphi(\bar\sigma)|) - (\varphi(\sigma) - \nu|\varphi(\sigma)|)](c^*x + \varrho\varphi(\sigma))$$
$$\leqq (\varphi(\sigma) - \nu|\varphi(\sigma)|)(c^*x + \varrho\varphi(\sigma)) + L(1 + \nu)(c^*x + \varrho\varphi(\sigma))^2$$

with $\bar\sigma$ meaning a value between σ and $\sigma + c^*x + \varrho\varphi(\sigma)$. Combining the last inequality and (11) we immediately get

$$V(F(z)) - V(z) + \vartheta\nu|\varphi(\sigma)|(c^*x + \varrho\varphi(\sigma)) - \vartheta L(1 + \nu)(c^*x + \varrho\varphi(\sigma))^2$$
$$+ \varepsilon(c^*x + \varrho\varphi(\sigma))^2 + \delta\varphi(\sigma)^2 \leqq 0 \qquad (12)$$

for all $(x,\sigma) \in \mathbb{R}^n \times S^1$.

Because of the assumption (ii) there exist some positive numbers ε_0 and δ_0 such that for all $(x,\xi) \in \mathbb{R}^n \times \mathbb{R}$

$$(\varepsilon - \vartheta L(1 + \nu))(c^*x + \varrho\xi)^2 + \vartheta\nu|\xi|(c^*x + \varrho\xi) + \delta\xi^2 \geqq \varepsilon_0(c^*x + \varrho\xi)^2 + \delta_0\xi^2$$

holds. Using that along with (12) we get the inequality

$$V(F(z)) - V(z) \leqq - \varepsilon_0(c^*x + \varrho\varphi(\sigma))^2 - \delta_0\varphi(\sigma)^2 \qquad (13)$$

for all $(x,\sigma) \in \mathbb{R}^n \times S^1$.

The conditions of Theorem 2 are fulfilled if for α we choose the function $\delta_0\varphi^2(\sigma)$. We have to show that this function satisfies (ii) of this theorem.

It follows from

$$\sum_{k=0}^{\infty} \varphi^2(\sigma(k)) < \infty$$

that $\varphi(\sigma(k)) \to 0$ for $k \to \infty$. Since by assumption the zeros of φ are isolated and $\sigma(k)$ satisfies the second relation of (9) we conclude, that $\sigma(k)$ converges to a zero of φ. Then the first part of the system (9) shows the convergence of $x(k)$ for $k \to \infty$ to zero.

Clearly, all conditions of Theorem 2 are satisfied. Recalling that A is a stable matrix and φ is bounded we get the boundedness of all semi-trajectories on $\mathbb{R}^n \times S^1$ and thus the implication of Theorem 4.

Literature

[1] Arapostathis, A., S.S. Sastry, and P. Varaiya: Global analysis of swing dynamics. IEEE Trans. on Circuits and Systems 29 (1982) 10, 673 - 679.

[2] Bakaev, Ju.N. and A.A. Guž: Optimal reception of signals of frequency modulation under the Doppler effect. Radiotekhnika i Elektronika 10 (1965) 1, 175 - 196 [Russian].

[3] Belykh, V.N. and V.I. Nekorkin: On qualitative investigation of multidimensional phase systems. Sibirskii Math. J. 18 (1977) 4, 723 - 735 [Russian].

[4] Brockett, R.W.: On the asymptotic properties of solutions of differential equations with multiple equilibria. J. of Diff. Equat. 44 (1982), 249 - 262.

[5] Gelig, A.Kh., G.A. Leonov and V.A. Jakubovič: Stability of systems with nonunique equilibrium state. Nauka, Moscow 1978 [Russian].

[6] Jakubovič, V.A.: The frequency theorem in control theory. Sibirsk. Math. J. 14 (1973) 2, 384 - 420 [Russian].

[7] Kobayashi, Sh. and K. Nomizu: Foundations of differential geometry. v. 1, Nauka, Moscow 1981 [Russian].

[8] Korjakin, Ju.A.: The Bakaev-Guž technique for discrete-time phase systems. In the book: Nonlinear oscillations and control theory. v. 1, Ishevsk, 1977 [Russian].

[9] Korjakin, Ju.A. and G.A. Leonov: The Bakaev-Guž technique for systems with multiple angular coordinates. Izvestija Akad. Nauk Kazakhskoi SSR, 3 (1976) [Russian].

[10] Korjakin, Ju.A., G.A. Leonov and V. Reitmann: Konvergenz im Mittel von Phasensystemen. ZAMM 58 (1978) 10, 435 - 441.

[11] Leonov, G.A.: The reduction theorem for nonstationary non-linearities. Vestnik LGU 7 (1978), 38 - 42 [Russian].

[12] Leonov, G.A., V. Reitmann and V.B. Smirnova: On the global asymptotic stability of ordinary and functional differential equations with multiple equilibria. Preprint 07-01-89, TU Dresden, 1989.

[13] Leonov, G.A. and V.B. Smirnova: Asymptotic behaviour of the solutions of integrodifferential equations with periodic non-linearities. Sibirsk. Math. J. 19 (1978) 6, 1406 - 1412 [Russian].

[14] Levi, M., F.C. Hoppensteadt and W.L. Miranker: Dynamics of the Josephson junction. Quarterly of Appl. Math. (1978) 7, 167 - 188.

[15] Noldus, E.J.: On the stability of systems having several equilibrium states. Appl. Sci. Res. 21 (1969), 218 - 233.

[16] Reitmann, V.: Globale Stabilität und Umlauflösungen für ein System zweier gekoppelter Josephson-Kontakt-Gleichungen. ZAMM (to be published).

[17] Shakhgildjan, V.V. and L.N. Beljustina (Ed.): Phase synchronization. Svjaz, Moscow, 1975 [Russian].

[18] Smith, R.A.: Some applications of Hausdorff dimension inequalities for ordinary differential equations. Proc. Roy. Soc. Edinburgh A 104 (1986) 3, 235 - 259.

ON THE LIMITS OF PERIODS OF CLOSED TRAJECTORIES CONTRACTING TO AN EQUILIBRIUM POINT

K.R. Schneider[1] and B. Wegner[2]

1. Introduction. A lot of papers and books have been devoted to the problem of bifurcation of time-periodic solutions from an equilibrium point of a system of ordinary or parabolic differential equations depending on a parameter [1,2,5,6,7,9,11 - 13]. In general, the problem under consideration is to obtain conditions guaranteeing the bifurcation of a periodic solution from an equilibrium point when μ crosses a critical value μ_o and if the limit of the corresponding period $\omega(\mu)$ as $\mu \longrightarrow \mu_o$ is given. This problem is called Hopf bifurcation. A main reason for the actual interest in Hopf bifurcation consists in the enormous importance of dynamical systems as mathematical models for biological, chemical, ecological and other processes.

In contrast with harmonic oscillations of periodic differential equations the period of limit cycles of autonomous systems is a-priori unknown. Throughout the following we call the smallest positive period of such oscillations the primitive period. Our paper is devoted to the following inverse problem in frame of Hopf-bifurcation theory: Given a family $\{\pi(t,\mu)\}$ of $\omega(\mu)$-periodic solutions of an autonomous differential system such that $\pi(t,\mu) \longrightarrow \pi(\mu_o)$ as $\mu \longrightarrow \mu_o$ where $\pi(\mu_o)$ is an equilibrium point. What can we say about the behaviour of $\omega(\mu)$ as $\mu \longrightarrow \mu_o$?

Concerning ordinary differential equations, this problem has been studied by the authors in [13,16,17] . In this article we continue these investigations and generalize the results to a class of semilinear differential equations in a Banach space. The crucial tool for establishing corresponding results for differential equations in Banach spaces is the method of integral manifolds [6,11,14].

2. Preliminaries. We consider in a real Banach space X with norm $\|.\|$ the differential equation

$$\frac{dx}{dt} + Ax = f(x,\mu) \tag{2.1}$$

depending on a real parameter $\mu \in M_o := \{\mu \in R: \ |\mu| < \mu_o\}$. Before we

[1] Karl-Weierstraß-Institut für Mathematik, DDR 1086 Berlin, Mohrenstra. 39

[2] Städt. Krankenhaus im Friedrichshain, DDR 1017 Berlin, Leninallee 49

formulate our assumptions on A and f let us introduce some notation and definitions.

Definition 2.1. We call a linear operator $L : X \longrightarrow X$ a sectorial operator if it is a closed densely defined operator such that, for some φ in $(0,\pi/2)$ and some $m \geq 1$ and real a, the sector

$$S_{a,\varphi} := \{z \in \mathbb{C} : \varphi \leq |\arg (z-a)| \leq \pi \ , \ z \neq a \}$$

is in the resolvent set $\rho(L)$ of L and

$$\|(L - zI)^{-1}\| \leq \frac{m}{|z-a|} \qquad \text{for all } z \in S_{a,\varphi} .$$

Remark 2.2. If L is a bounded linear operator on X or if L is a selfadjoint densely defined operator in a Hilbert space and is bounded below, then L is sectorial [6]. Many elliptic boundary value problems define sectorial operators [4].

Theorem 2.3. [6] If L is a sectorial operator, then -L is the infinitesimal generator of an analytic semigroup $\{e^{-Lt}\}_{t\geq 0}$, where e^{-Lt} is defined by the Dunford-Taylor integral

$$e^{-Lt} = \frac{1}{2\pi i} \int_{\Gamma} (L + zI)^{-1} e^{zt} \, dz, \quad t \geq 0, \tag{2.2}$$

here Γ is a contour in $\rho(-L)$ with $\arg z \longrightarrow \pm \theta$ as $|z| \longrightarrow \infty$ for some θ in $(\pi/2,\pi)$.

If L is a sectorial operator in X, then there is a real number $\tilde{l}$ such that the spectrum $\sigma(\tilde{L})$ of the operator $\tilde{L} := L + \tilde{l}I$ satisfies Re $\sigma(\tilde{L}) > 0$. For the operator $\tilde{L}$ the fractional powers $\tilde{L}^{\alpha}$, α any real number, are defined [6]. The domain of $\tilde{L}^{\alpha}$ is denoted by X^{α}_{L} .

Theorem 2.4. [6] Let $L: X \longrightarrow X$ be a sectorial operator. Then X^{α}_{L} is a Banach space with the norm $\|\cdot\|_{\alpha}$ for $\alpha \geq 0$, where $\|x\|_{\alpha} := \|\tilde{L}^{\alpha}x\|$ for $x \in X^{\alpha}_{L}$; $X^{0}_{L} = X$, and for $\alpha \geq \beta \geq 0$ X^{α}_{L} is a dense subspace of X^{β}_{L} with continuous inclusion. (Different choices of $\tilde{l}$ give equivalent norms on X^{α}_{L} , so we supress the dependence on the choice of $\tilde{l}$.)

Definition 2.5. If $L: X \longrightarrow X$ is linear, then a set $\sigma \subseteq \tilde{\sigma}(L) := \sigma(L) \cup \{\infty\}$ is said to be a spectral set if both σ and $\tilde{\sigma}(L)\backslash\sigma$ are closed in the extended plane $\mathbb{C} \cup \{\infty\}$.

Let $L: X \longrightarrow X$ be a closed linear operator. We suppose that $\tilde{\sigma}_1$ is a nonempty bounded spectral set of L. Let $\tilde{P}_1$ be the projection associated with $\tilde{\sigma}_1$. It is well-known [8] that $\tilde{P}_1$ is defined by

$$\tilde{P}_1 := -\frac{1}{2\pi i}\int_{\tilde{\Gamma}_1} (L - zI)^{-1}\,dz \tag{2.3}$$

where $\tilde{\Gamma}_1$ is a closed contour in $\rho(L)$ containing $\tilde{\sigma}_1$ in its interior. We set $\tilde{P}_2 := I - \tilde{P}_1$, $\tilde{X}_j := \tilde{P}_j X$, $\tilde{L}_j := L|\tilde{X}_j$, $j = 1,2$.

Theorem 2.6. [6] Let $L: X \longrightarrow X$ be a closed linear operator, let $\tilde{\sigma}_1$ be a nonempty bounded spectral set of L. Then we have

(i). $X = \tilde{X}_1 \oplus \tilde{X}_2$.

(ii). $\tilde{L}_1$ is bounded on $\tilde{X}_1$.

(iii). $\tilde{L}_j\tilde{X}_j \subset \tilde{X}_j$, $\sigma(\tilde{L}_j) = \tilde{\sigma}_j$ for $j = 1,2$.

If we additionally assume that L is a sectorial operator, then there are positive constants $\tilde{c}_1$, $\tilde{c}_2$ such that

(iv). If Re $\sigma(\tilde{L}_1) < \delta$, then

$$\|e^{-\tilde{L}_1 t}\| \le \tilde{c}_1 e^{-\delta t} \text{ for } t \le 0. \tag{2.4}$$

(v). If Re $\sigma(\tilde{L}_2) > \beta > 0$, then for $t > 0$, $\alpha \ge 0$

$$\left\|e^{-\tilde{L}_2 t}\right\| \le \tilde{c}_2 e^{-\beta t}, \quad \left\|\tilde{L}_2^{\alpha}\, e^{-\tilde{L}_2 t}\right\| \le \tilde{c}_{\alpha}\, t^{-\alpha}\, e^{-\beta t} \tag{2.5}$$

where $c_{\alpha} > 0$ is bounded for α on any compact interval of $[0,\infty)$.

Remark 2.7. If L has a compact resolvent, then X_1 is finite dimensional [8].

In what follows we suppose on the operator A in (2.1):

(A_1). $A : X \longrightarrow X$ is sectorial.

Let $\tilde{a}$ be such a number that Re $\sigma(A + \tilde{a}I) > 0$. We denote by X^{α} the domain of $\tilde{A}^{\alpha} := (A + \tilde{a}I)^{\alpha}$. Let $\mathscr{G}^{\alpha}$ be a neighbourhood of the origin in X^{α}. Let $S_{\varepsilon}(0) := \{x \in \mathscr{G}^{\alpha}: \|x\|_{\alpha} \le \varepsilon\}$, let $\mathscr{L}(Y,Z)$ be the space of bounded linear operators defined on Y and mapping into Z.

$(F_{1,1})$. There is an α, $0 \le \alpha < 1$, such that $f: \mathscr{G}^{\alpha} \times M_0 \longrightarrow X$ is continuous, there are an $\varepsilon_0 > 0$ with $S_{\varepsilon_0}(0) \subseteq \mathscr{G}^{\alpha}$ and a function l:

$[0,\varepsilon_o] \longrightarrow R^+$ with $l(\varepsilon) \longrightarrow 0$ as $\varepsilon \longrightarrow 0$ such that

$$\|f(x^{(1)},\mu) - f(x^{(2)},\mu)\| \le l(\varepsilon)\ \|x^{(1)} - x^{(2)}\|_\alpha$$

$$\forall\ x^{(1)},\ x^{(2)} \in S_\varepsilon(0),\ 0 \le \varepsilon \le \varepsilon_o,\ \forall \mu \in M_o .$$

Definition 2.8. By a local solution $\bar{x}(t,t_o,x,\mu)$ of the Cauchy problem

$$\frac{dx}{dt} + Ax = f(x,\mu),\ x(t_o) = x \in X \tag{2.6}$$

we mean a continuous function $\bar{x} : [t_o,t_1) \longrightarrow X$, $t_1 > t_o$, with $\bar{x}(t_o;t_o,x,\mu) = x$ such that for $t \in (t_o,t_1)$ we have $\bar{x}(t;t_o,x,\mu) \in \mathscr{G}^\alpha \cap$ domain of A, $d\bar{x}/dt$ exists and (2.1) is satisfied.

A solution of (2.6) is said to be global whenever $t_1 = +\infty$. We call a global solution $\bar{x}(t;t_o,x,\mu)$ of (2.6) a periodic solution if there is a number $\omega \ne 0$ such that $\bar{x}(t;t_o,x,\mu) = \bar{x}(t+\omega;t_o,x,\mu)\ \forall\ t \in R$. A periodic solution of (2.1) satisfying $\bar{x}(t;t_o,x,\mu) = x\ \forall\ t \in R$ is said to be an equilibrium point.

Theorem 2.9. [14] Assume $(A_1),(F_{1,1})$ hold. Then a local solution of (2.6) satisfies on (t_o,t_1)

$$\tilde{x}(t,x,\mu) = e^{-A(t-t_o)}x + \int_{t_o}^{t} e^{-A(t-s)}f(\tilde{x}(s,x,\mu),\mu)ds. \tag{2.7}$$

Conversely, a continuous function $\bar{x} : (t_o,t_1) \longrightarrow X^\alpha$ obeing (2.7) is a local solution of (2.6) on (t_o,t_1).

We impose now on A a second assumption:

(A_2). The set σ_o, $\sigma_o := \{ \lambda \in \sigma(A): \operatorname{Re} \lambda = 0 \}$ is a nonempty spectral set.

Since A is a sectorial operator the spectral set σ_o is bounded.
Let

$$\sigma_1 := \{\lambda \in \sigma(A): \operatorname{Re} \lambda < 0\},$$
$$\sigma_2 := \{\lambda \in \sigma(A): \operatorname{Re} \lambda > 0\}.$$

From the assumptions (A_1) and (A_2) it follows that σ_1 and σ_2 are spectral sets where σ_1 is bounded. Hence, there is a number $\beta > 0$ such that $\operatorname{Re} \sigma_2 > \beta$, $\operatorname{Re} \sigma_1 < -\beta$. Let us denote by P_j, $j = 0,1,2$, the projection associated with σ_j where P_j is defined according to (2.3) by

$$P_j := -\frac{1}{2\pi i}\int_{\Gamma_j} (A - zI)^{-1}\, dz \tag{2.8}$$

where Γ_j is a closed contour in $\rho(A)$ containing σ_j in its interior.

Let $X_j := P_jX$, $A_j := A|X_j$, $j = 0,1,2$. Then we have the following relations according to Theorem 2,6:

(i). $X = X_0 \oplus X_1 \oplus X_2$, $A_jX_j \subset X_j$, $\sigma(A_j) = \sigma_j$ for $j = 0,1,2$.

(ii). A_0 is a bounded linear operator satisfying

$$\left\| e^{-A_0 t} \right\| \le c_0 \; e^{\beta|t|/2} \qquad \text{for } t \in R,\; c_0 \ge 1 . \tag{2.9}$$

(iii). $A_1: X_1 \longrightarrow X_1$ is a bounded linear operator such that for $t \le 0$ and $c_1 \ge 1$

$$\left\| e^{-A_1 t} \right\| \le c_1 \; e^{-\beta t} \qquad \text{for } t \le 0,\; c_1 \ge 1 . \tag{2.10}$$

(iv). A_2: $X_2 \longrightarrow X_2$ is a sectorial operator satisfying for $\alpha \ge 0$, $t > 0$

$$\left\| A_2^{\alpha} \; e^{-A_2 t} \right\| \le \; c_\alpha \; t^{-\alpha} \; e^{-\beta t} \tag{2.11}$$

where $c_\alpha > 0$ is bounded for α on any compact interval of $[0,\infty)$.

In what follows we denote by $\|\cdot\|_i$ the norm in X_i, $i = 0,1,2$, induced by the norm in X, and by $\|\cdot\|_{2,\alpha}$ with $\|x\|_{2,\alpha} := \|A_2^{\alpha}x\|_2$ for $x \in X_2^{\alpha}$ the norm in X_2^{α} . In $X_0 \times X_1 \times X_2^{\alpha}$ we introduce a norm $\||\cdot|\|$ by

$$\||(y_0,y_1,y_2)|\| := \|y_0\| + \|y_1\|_1 + \|y_2\|_{2,\alpha} .$$

Using the notation $x = (x_0,x_1,x_2)$, $x_0 \in X_0, x_1 \in X_1$, $x_2 \in X_2^{\alpha}$ we may rewrite the differential equation (2.1) in the equivalent form

$$\begin{aligned} \frac{dx_0}{dt} + A_0 &= P_0 f(x_0,x_1,x_2,\mu), \\ \frac{dx_1}{dt} + A_1 &= P_1 f(x_0,x_1,x_2,\mu), \\ \frac{dx_2}{dt} + A_2 &= P_2 f(x_0,x_1,x_2,\mu), \end{aligned} \tag{2.12}$$

moreover we can reformulate the assumption $(F_{1,1})$:

(F_1): There is an α, $0 \le \alpha < 1$, such that $f: \mathcal{G}_0 \times \mathcal{G}_1 \times \mathcal{G}_2^{\alpha} \times M_0 \longrightarrow X$ is continuous where $\mathcal{G}_i$, $i = 1,2$, is a neighbourhood of the origin in X_i, $\mathcal{G}_2^{\alpha}$ is a neighbourhood of the origin in X_2^{α}. There is a function $l(\epsilon)$ with

$l(\varepsilon) \longrightarrow 0$ as $\varepsilon \longrightarrow 0$ such that

$$\|f(x_0^{(1)},x_1^{(1)},x_2^{(1)}) - f(x_0^{(2)},x_1^{(2)},x_2^{(2)})\| \le$$

$$\le l(\varepsilon)\ |\|(x_0^{(1)},x_1^{(1)},x_2^{(1)}) - (x_0^{(2)},x_1^{(2)},x_2^{(2)})\||$$

$$\forall\ (x_0^{(1)},x_1^{(1)},x_2^{(1)}),\ (x_0^{(2)},x_1^{(2)},x_2^{(2)}) \in \mathcal{G}_0 \times \mathcal{G}_1 \times \mathcal{G}_2^{\alpha} \quad \text{satisfying}$$

$$|\|(x_0^{(1)},x_1^{(1)},x_2^{(1)})\|| \le \varepsilon,\ |\|(x_0^{(2)},x_1^{(2)},x_2^{(2)})\|| \le \varepsilon,\ 0 \le \varepsilon \le \varepsilon_0.$$

Since we are interested in the behaviour of the minimal period of a periodic solution of (2.1) contracting to an equilibrium point as $\mu \longrightarrow \mu_0$ we further suppose :

(F_2). $f(0,\mu) \equiv 0 \quad \forall\ \mu \in M_0$. (2.13)

(B_1). There exists a sequence $\{\mu_n\} \subseteq M_0$ with

$$\lim_{n \longrightarrow \infty} \mu_n = 0 \tag{2.14}$$

such that for each μ_n the equation (2.1) has in $\mathcal{G}_0 \times \mathcal{G}_1 \times \mathcal{G}_2^{\alpha}$ a non-constant periodic solution $\pi_n(t)$ satisfying

$$\lim_{n \longrightarrow \infty} \||\pi_n|\|_C = 0 \tag{2.15}$$

where $\|u\|_C := \max_{t \in [0,T]} \|u(t)\|$ for a continuous mapping $u\colon [0,T] \longrightarrow Y$, Y is a Banach space with norm $\|.\|$.

Let ω_n be the primitive period of $p_n(t)$. Any a-priori estimate of ω_n is not known. To investigate the asymptotic behaviour of ω_n as $n \longrightarrow \infty$ we use the method of integral manifolds.

<u>Definition 2.10</u>. A set $\mathcal{M} \in X_0 \times X_1 \times X_2^{\alpha}$ is said to be a local integral manifold of the system (2.12) iff to any $x \in \mathcal{M}$ there is an interval (t_0,t_1) such that the solution $\bar{x}(t;t_0,x,\mu)$ of (2.12) exists on (t_0,t_1) and belongs to $\mathcal{M}$.

From [14] we obtain the following result.

<u>Theorem 2.11</u>. Assume (A_1), (A_2), (F_1), (F_2) hold. Then there exists a function $\zeta\colon \tilde{\mathcal{G}}_0 \times M_1 \longrightarrow X_1 \times X_2^{\alpha}$ where $\tilde{\mathcal{G}}_0 \subseteq \mathcal{G}_0$ and $0 \in M_1 \subseteq M_0$ such that

(i). $(x_1,x_2) = \zeta(x_0,\mu)$ is a local integral manifold $\mathcal{M}_\mu$ of the system (2.12).

(ii). ζ is continuous and satisfies

$$\|\zeta(x_0^{(1)},\mu) - \zeta(x_0^{(2)},\mu)\|_{1,2} \le \varkappa(\varepsilon)\ \|x_0^{(1)} - x_0^{(2)}\|_0 \qquad (2.16)$$

$\forall\ \mu_0 \in M_1,\ x_0^{(1)},\ x_0^{(2)} \in \tilde{\mathcal{G}}_0,\ \|x_0^{(1)}\| \le \varepsilon,\ \|x_0^{(2)}\| \le \varepsilon$ where $\varkappa(\varepsilon) \longrightarrow 0$ as $\varepsilon \longrightarrow 0$ and where $\|\cdot\|_{1,2} := \|\cdot\|_1 + \|\cdot\|_{2,\alpha}$.

(iii). ζ is uniformly bounded on $\tilde{\mathcal{G}}_0 \times M_1$ and obeys

$$\zeta(0,\mu) \equiv 0\ \forall\ \mu \in M_1. \qquad (2.17)$$

(iv). Any global solution $\bar{x}(t,\mu) = (\bar{x}_0(t,\mu),\bar{x}_1(t,\mu),\bar{x}_2(t,\mu))$ of (2.1) such that $\sup_{t\in\mathbb{R}} \left[\|x_1\|_1 + \|x_2\|_{2,\alpha}\right]$ is sufficiently small is located on $\mathcal{M}_\mu$.

It follows from Theorem 2.11 that all periodic solutions of (2.1) lying in a small neighbourhood of the origin are situated on this integral manifold. Thus, Theorem 2.11 provides the possibility to reduce the investigation of the closed orbits of the system (2.1) near the origin to the investigation of the periodic solutions of the system

$$\frac{dx_0}{dt} + A_0 = P_0 f(x_0,\zeta(x_0,\mu),\mu) := f_0(x_0,\mu) \qquad (2.18)$$

where A_0 is a bounded linear operator.

To be able to assure the existence of a smooth integral manifold of the system (2.12) we have to strengthen hypothesis (F_1), furthermore we need an additional assumption on the Banach space $X_0 \times X_1 \times X_2^\alpha$. To this end we define:

<u>Definition 2.12.</u> We say the Banach space Y has the property $(P)_l$ if there exists a function $\rho \in C^l(Y,[0,\infty))$ such that $\rho \equiv 0$ for $\|y\| \ge \delta$ and $\rho \equiv 1$ for $\|y\| \le \delta/2$, for some $\delta > 0$.

<u>Remark 2.13.</u> It is well-known that if X is finite dimensional, or if X is a real Hilbert space then it has the property $(P)_\infty$.

(S_1). $X_0 \times X_1 \times X_2^\alpha$ is a Banach space with the property $(P)_1$.

$(\tilde{F}_1)$. There is an α, $0 \le \alpha < 1$, such that $f: \mathcal{G}_0 \times \mathcal{G}_1 \times \mathcal{G}_2^\alpha \times M_0 \longrightarrow X$ is continuous and differentiable with respect to x such that $f_x: \mathcal{G}_0 \times \mathcal{G}_1 \times \mathcal{G}_2^\alpha \times M_0 \longrightarrow \mathcal{L}(X_0 \times X_1 \times X_2^\alpha)$ is uniformly continuous and satisfies

$$f_x(0,0) = 0. \qquad (2.19)$$

Theorem 2.14.[11] Assume (A_1), (A_2), $(\tilde{F}_1)$, (F_2), (S_1) hold. Then the assertions of Theorem 2.11 are valid. Additionally we have the following smoothness property: ζ is differentiable with respect to x_o on $\tilde{\mathscr{G}}_o \times M_1$ where $\zeta_{x_o}(x_o,\mu)$ is uniformly continuous in (x_o,μ) and satisfies

$$\zeta_{x_o}(0,0) = 0. \tag{2.20}$$

3. Formulation of the problems under consideration

By assumption (B_1) there exists a sequence $\{\pi_n(t)\}$ of non-constant periodic solutions of (2.1) having the primitive period ω_n and contracting to the origin as $n \longrightarrow \infty$. We are interested in the asymptotic behaviour of the sequence $\{\omega_n\}$. The assumption (B_1) does not exclude that the elements of the sequence $\{\pi_n(t)\}$ belongs to different families of periodic solutions shrinking to $x = 0$ as $n \longrightarrow \infty$. Thus, the sequence $\{\omega_n\}$ may contain more than one cluster point. In the sequel we consider the following problems.

Problem P_1. Assume (A_1), (A_2), (F_1), (F_2), (B_1) to be valid. Additionally we suppose

(B_2). There is a subsequence $\{\bar{\mu}_n\}$ of $\{\mu_n\}$ such that the corresponding sequence of primitive periods $\{\bar{\omega}_n\}$ satisfies

$$\lim_{n \longrightarrow \infty} \bar{\omega}_n = \bar{\omega}, \quad 0 < \bar{\omega} < +\infty .$$

What can we say about $\bar{\omega}$ under these assumptions?

Próblem P_2. Suppose the hypotheses (A_1), (A_2), (F_1), (F_2), (B_1) hold. Under which conditions do there are subsequences $\{\tilde{\mu}_n\}$ and $\{\hat{\mu}_n\}$ of $\{\mu_n\}$ such that for the corresponding sequences of primitive periods $\{\tilde{\omega}_n\}$ and $\{\hat{\omega}_n\}$

$$\lim_{n \longrightarrow \infty} \tilde{\omega}_n^{-1} = 0 , \qquad \lim_{n \longrightarrow \infty} \hat{\omega}_n = 0.$$

Using the results of the preceding section it sufficies to consider the corresponding periodic solutions of the reduced system (2.18) where A_o is a bounded linear operator. Let us assume that to the sequences $\{\bar{\omega}_n\},\{\tilde{\omega}_n\},\{\hat{\omega}_n\}$ do correspond the sequences $\{\bar{p}_n(t)\},\{\tilde{p}_n(t)\},\{\hat{p}_n(t)\}$ of periodic solutions of the system (2.18).
Substituting $t = \omega\tau$ in (2.18) we get

$$\frac{dx_o}{d\tau} + \omega A_o x_o = \omega f_o(x_o,\mu). \tag{3.1}$$

The following lemma is obvious.

Lemma 3.1. Assume (A_1), (A_2), (F_1) are valid. Then the initial value problem

$$\frac{dx_o}{d\tau} + \omega A_o x_o = \omega f_o(x_o,\mu), \quad x_o(0) = x_o^o \tag{3.2}$$

is equivalent to the integral equation

$$x_o(\tau,\mu) = e^{-A_o\omega\tau} x_o^o + \omega \int_o^{\tau} e^{-A_o\omega(\tau-\sigma)} f_o(x_o(\sigma,\mu),\mu)\, d\sigma . \tag{3.3}$$

In the sequel we use this lemma to treat the posed questions.

4. Investigation of Problem P_1

By the hypotheses (B_1) and (B_2) there is a subsequence $\{\bar{\omega}_n\}$ of $\{\omega_n\}$ such that (3.1) has a 1-periodic solution $\bar{p}_n(t)$ for $\omega = \bar{\omega}_n$, $\mu = \mu_n$ contracting to the origin as $n \longrightarrow \infty$ where

$$\lim_{n \longrightarrow \infty} \bar{\omega}_n = \bar{\omega} , \quad 0 < \bar{\omega} < \infty .$$

Let us first consider problem P_1 under the stronger hypothesis $(\tilde{F}_1)$ and by assuming

(A_3). A_o is invertible.

Then we have:

Theorem 4.1. Suppose (A_1) - (A_3), $(\tilde{F}_1)$, (F_2), (B_1), (B_2) hold. Then there is a pair of imaginary eigenvalues $\pm$ ib, $b > 0$ in $\sigma(A_o)$ and a natural number m such that

$$\bar{\omega} = \frac{2\pi}{b} m . \tag{4.1}$$

Proof. Let us denote by $C_1^k(R,X_o)$ the space of 1-periodic functions h: $R \longrightarrow X_o$ which are k-times continuously differentiable, further we set

$$\mathcal{F}(\omega,\mu,x_o) := \frac{dx_o}{d\tau} + \omega A_o x_o - \omega f_o(x_o,\mu) \tag{4.2}$$

such that $\mathcal{F}$ maps $R \times M_1 \times C_1^1(R,\tilde{\mathcal{X}}_o)$ into $C_1(R,X_o)$. Under our hypotheses $\mathcal{F}$ is continuously differentiable. Moreover we have

$$\mathcal{F}(\omega,\mu,0) = 0 \quad \text{for } (\omega,\mu) \in R \times M_1 .$$

We are interested in nontrivial solutions of $\mathcal{F} = 0$ in a neighbourhood of $(\bar{\omega},0,0)$. Let us denote by $\mathcal{F}_{x_o}(\bar{\omega},0,0)$ the partial Fréchet derivative of the map $\mathcal{F}$ with respect to x_o at $(\omega,\mu,x_o) = (\bar{\omega},0,0)$. If $\mathcal{F}_{x_o}(\bar{\omega},0,0)$ is an isomorphism between $C_1^1(R,X_o)$ and $C_1(R,X_o)$, then the implicite function theorem implies that for the zeroes (ω,μ,x_o) of $\mathcal{F}$ near $(\bar{\omega},0,0)$ x_o is given uniquely as a function of (ω,μ). But such a function is known, namely $x_o \equiv 0 \ \forall \ (\omega,\mu) \in R \times M_1$. Consequently, in this case there are no nontrivial 1-periodic solutions of (3.1) with ω near $\bar{\omega}$ which is close to (0,0). From (4.2), (2.19) - (2.20) it follows

$$\mathcal{F}_{x_o}(\bar{\omega},0,0)v_o = \frac{dv_o}{d\tau} + \bar{\omega}\, A_o v_o \ .$$

Thus, $\mathcal{F}_{x_o}(\bar{\omega},0,0)$ is singular if and only if there is a solution of

$$\frac{dv_o}{d\tau} + \bar{\omega}\, A_o v_o = 0 \ , \ v_o \in C_1^1(R,X_o)\setminus\{0\} \ . \tag{4.3}$$

By using Lemma 3.1 the problem (4.3) is equivalent to the existence of an element $v_o^o \neq 0$, $v_o^o \in X_o$ such that

$$v_o^o = e^{-A_o\bar{\omega}}\, v_o^o \ .$$

This relation implies that $e^{-A_o\bar{\omega}}$ has the eigenvalue +1. By the spectral mapping theorem [3], we have

$$\sigma\left(e^{-A_o\bar{\omega}} \right) = e^{-\bar{\omega}}\, \sigma(A_o).$$

Thus, since A_o is invertible, there is a pair of imaginary eigenvalues $\pm ib$, $b > 0$, of the operator A_o such that $e^{\pm ib\bar{\omega}} = 1$. Therefore, $\bar{\omega}$ satisfies the relation (4.1), q.e.d.

The assumption $(\tilde{F}_1)$ does not belong to the hypotheses of Problem P_1. Thus, the implicite function theorem does not apply. To obtain the same result as in Theorem 4.1 we impose an additional assumption on the space X_o.

(S_2). The space X_o has finite dimension.

<u>Theorem 4.2.</u> Suppose the hypotheses (A_1) - (A_3), (F_1), (F_2), (S_2), (B_1), (B_2) hold. Then the assertion of Theorem 4.1 is valid.

<u>Proof.</u> We set $q_n(\tau) := p_n(\bar{\omega}_n\tau)$. Thus, $q_n(\tau)$ is a 1-periodic solution of (3.1) for $\mu = \mu_n$ and satisfies by (3.3)

$$q_n(\tau) = e^{-A_o\bar{\omega}_n\tau} q_n(0)+\bar{\omega}_n \int_o^\tau e^{-A_o\bar{\omega}_n(\tau-\sigma)} f_o(q_n(\sigma),\mu_n)\, d\sigma. \tag{4.4}$$

Since (3.2) is an autonomous equation, we may suppose without loss of generality

$$\|q_n(0)\|_o = \|q_n\|_{o,c}.$$

Setting

$$v_n(\tau) := q_n(\tau)/\|q_n\|_{o,c} \tag{4.5}$$

and taking into account $q_n(0) = q_n(1)$ we obtain from (4.4)

$$v_n(0) = e^{-A_o\bar{\omega}_n} v_n(0) + \bar{\omega}_n \int_o^1 e^{-A_o\bar{\omega}_n(1-\sigma)} \frac{f_o(q_n(\sigma),\mu_n)}{\|q_n\|_{o,c}}\, d\sigma. \tag{4.6}$$

From the hypotheses (F_1), (F_2), and by Theorem 2.11 we get

$$\|f_o(x_o,\mu)\|_o/\|x_o\| \longrightarrow 0 \text{ as } \|x_o\| \longrightarrow 0.$$

Thus, by means of the boundedness of $e^{A_o t}$ and by (B_2) it follows immediately

$$\lim_{n \longrightarrow \infty} \int_o^1 e^{-A_o\bar{\omega}_n(1-\sigma)} \frac{f_o(q_n(\sigma),\mu_n)}{\|q_n\|_{o,c}}\, d\sigma = 0 . \tag{4.7}$$

According to (S_2) the sphere $S(0,1) := \{x_o \in X_o,\ \|x_o\|_o = 1 \}$ is compact. Thus, there is a convergent subsequence $\{v_{n_k}(0)\}$ of $\{v_n(0)\}$ with $v_{n_k}(0) \longrightarrow v_o \in S(0,1)$ as $n \longrightarrow \infty$. Using the boundedness of A_o and the hypothesis (B_2) it can be easily seen that

$$\lim_{n_k \longrightarrow \infty} e^{-A_o\bar{\omega}_{n_k}} v_{n_k}(0) = e^{-A_o\bar{\omega}} v_o.$$

Applying this relation and (4.7) we get from (4.6)

$$v_o = e^{-A_o\bar{\omega}} v_o, \quad \|v_o\| = 1 . \tag{4.8}$$

By the same way as in Theorem 4.1 we can complete the proof.

Corollary 4.3. The assertion of Theorem 4.2 remains valid if we replace the hypothesis (A_3) by the weaker condition: $A_o v_o \neq 0$ where v_o is the limit of a convergent subsequence of $\{v_n(0)\}$ where $\{v_n(\tau)\}$ is defined in (4.5)

Proof. From (4.8) and $A_1 v_o \neq 0$ we get that the function $v(\tau) := e^{-A_o \bar{\omega}\tau} v_o$ is a non-constant 1-periodic solution of the linear equation $\frac{dv}{dt} = -A_o \bar{\omega} v$ in the finite dimensional space X_o. Taking into account the relations $\bar{\omega} > 0$ and $A_o v_o \neq 0$ we can complete the proof in the same lines as above.

Remark 4.4. A further relaxation of the assumptions of Theorem 4.2 can be obtained if we replace (S_2) by the assumption that the sequence $\{v_n(0)\}$ contains a subsequence converging to an element v_o with $A_o v_o \neq 0$.

Remark 4.5. If we assume that the resolvent of A is compact, then the hypotheses (A_2) and (S_2) are valid. As mentioned in [6] the most classical boundary value problems generate semilinear equations where A has a compact resolvent.

Concerning Theorem 4.1 and Theorem 4.2 there arises the following question: Under which conditions can we put m = 1 in the relation (4.1)? The following theorem gives a positive answer in case that we can reduce (2.1) to a two-dimensional system of ordinary differential equations.

Theorem 4.6. Assume the hypotheses (A_1), (F_1), (F_2), (B_1),(B_2) hold. Additionally we suppose

(A_4). A has a compact resolvent where $\sigma(A) \cap \{\mathrm{Re}\ z = 0\}$ consists of the simple eigenvalues $\pm ib$, $b > 0$.

Then we have for any subsequence $\{\bar{\omega}_n\}$ of $\{\omega_n\}$ with $\bar{\omega}_n \longrightarrow \bar{\omega}$

$$\bar{\omega} = \frac{2\pi}{b}. \tag{4.9}$$

Proof. Under the assumptions above the system (2.18) is a finite dimensional system of ordinary differential equations where $\sigma(A) \cap \{\mathrm{Re}\ z\ 0\} = \{\pm ib,\ b > 0\}$. Using the center manifold theorem [5] we can reduce the system (3.1) to the following one

$$\frac{dw_1}{d\tau} = -\omega\, b\, w_2 + \omega\, h_1(w_1, w_2, \mu),$$
$$\frac{dw_2}{d\tau} = \omega\, b\, w_1 + \omega\, h_2(w_1, w_2, \mu). \tag{4.10}$$

Here $h(w,\mu) := (h_1(w_1,w_2,\mu), h_2(w_1,w_2,\mu))$ has the same regularity properties as $f(x,\mu)$ and satisfies

$$h(0,\mu) \equiv 0 \;\forall\; |\mu| \le \mu_1. \tag{4.11}$$

By $w_1 = r\cos\varphi$, $w_2 = r\sin\varphi$ we introduce polar coordinates. Then in a small neighbourhood of $w = 0$ the system (4.10) is equivalent to

$$\frac{dr}{d\tau} = \omega\left[\tilde{h}_1(r,\varphi,\mu)\cos\varphi + \tilde{h}_2(r,\varphi,\mu)\sin\varphi\right],$$
$$\frac{d\varphi}{d\tau} = \omega b\left[1 + \frac{\tilde{h}_2(r,\varphi,\mu)\cos\varphi - \tilde{h}_1(r,\varphi,\mu)\sin\varphi}{br}\right] \tag{4.12}$$

and to

$$\frac{dr}{d\varphi} = \frac{1}{b}\,\frac{\tilde{h}_1(r,\varphi,\mu)\cos\varphi + \tilde{h}_2(r,\varphi,\mu)\sin\varphi}{1 + \dfrac{\tilde{h}_2(r,\varphi,\mu)\cos\varphi - \tilde{h}_1(r,\varphi,\mu)\sin\varphi}{br}} \tag{4.13}$$

where $\tilde{h}_i(r,\varphi,\mu) := h_i(r\cos\varphi, r\sin\varphi)$, $i = 1,2$, satisfies $\tilde{h}_i(r,\varphi,\mu) = o(1)$ as $r \longrightarrow 0$ uniformly in φ and μ, $0 \le \varphi \le 2\pi$, $|\mu| \le \mu_2 < \mu_1$. Setting

$$h(r,\varphi,\mu) := \frac{\tilde{h}_2(r,\varphi,\mu)\cos\varphi - \tilde{h}_1(r,\varphi,\mu)\sin\varphi}{br} \tag{4.14}$$

we have by (4.11)

$$h(r,\varphi,\mu) = o(1) \text{ as } |\mu| + r \longrightarrow 0 \text{ uniformly in } \varphi. \tag{4.15}$$

Let $\{r_n(\varphi,\mu_n)\}$ be the sequence of 2π -periodic solutions of (4.13) corresponding to the sequence $\{\bar{p}_n(t)\}$ of $\bar{\omega}_n$ - periodic solutions of the system (2.18) contracting to the origin as $n \longrightarrow \infty$. Substituting these solutions into (4.12) and integrating the second equation over τ, $0 \le \tau \le 1$, we obtain

$$\bar{\omega}_n b = \int_0^{2\pi} \frac{d\varphi}{1 + h(r_n(\varphi,\mu_n),\varphi,\mu_n)} = 2\pi - \int_0^{2\pi} \frac{h(r_n(\varphi,\mu_n),\varphi,\mu_n)}{1 + h(r_n(\varphi,\mu_n),\varphi,\mu_n)}\, d\varphi. \tag{4.16}$$

By (4.15) and (H_1) we get from (4.16) $\bar{\omega}\, b = 2\pi$, q.e.d.

Remark 4.7. If $\sigma(A)$ has more than one pair of simple imaginary eigenvalues than we cannot put $m = 1$ in (4.1) for any subsequence $\{\bar{\omega}_n\}$ of $\{\omega_n\}$ converging to $\bar{\omega}$. To show this we consider the differential system

$$\begin{aligned}
\frac{dv_1}{dt} &= -av_2 + v_1(v_1^2 + v_2^2 - \mu^2)(v_3^2 + v_4^2 - \mu^2)\,,\\
\frac{dv_2}{dt} &= av_1 + v_2(v_1^2 + v_2^2 - \mu^2)(v_3^2 + v_4^2 - \mu^2)\,,\\
\frac{dv_2}{dt} &= -bv_4 + v_3(v_1^2 + v_2^2 - \mu^2)(v_3^2 + v_4^2 - \mu^2)\,,\\
\frac{dv_2}{dt} &= bv_3 + v_4(v_1^2 + v_2^2 - \mu^2)(v_3^2 + v_4^2 - \mu^2)\,.
\end{aligned} \tag{4.17}$$

The corresponding linear operator A_0 has the eigenvalues $\pm ia$, $\pm ib$. As we immedeately see the system (4.17) has the families of closed orbits

$$\Gamma_1 := \left\{v_1 = \mu_k \cos at,\ v_2 = \mu_k \sin at,\ v_3 = v_4 = 0\right\} \text{ with } \bar{\omega} = 2\pi/a,$$

$$\Gamma_2 := \left\{v_3 = \mu_k \cos bt,\ v_4 = \mu_k \sin bt,\ v_1 = v_2 = 0\right\} \text{ with } \bar{\omega} = 2\pi/b.$$

For these families we have $m = 1$ in (4.1). Assuming that a and b are commensurate we have further the family of closed orbits

$$\Gamma_3 := \left\{v_1 = \mu_k \cos at, v_2 = \mu_k \sin at,\ v_3 = \mu_k \cos bt, v_4 = \mu_k \sin bt\right\}.$$

If we have $a/b = m_a/m_b$ where m_a and m_b are natural numbers without a common divisor, then to the family Γ_3 does belong the sequence of primitive periods $\bar{\omega}_k = 2\pi m_b/a = 2\pi m_a/b$. Thus we have in (4.1) $m = m_a$ or $m = m_b$.

5. Investigation of the Problem P_2.

An important tool to treat the Problem P_2 is the following theorem due to A. Lasota and J.A. Yorke [10].

Theorem 5.1. Let X be a Banach space with norm $\|.\|$, let G be a subset in X, $h : G \longrightarrow X$ satisfies for any $z_1, z_2 \in G$ and some > 0

$$\|h(z_1) - h(z_2)\| \le \alpha \|z_1 - z_2\|\,. \tag{5.1}$$

Moreover we assume that the differential system

$$\frac{dz}{dt} = h(z) \qquad (5.2)$$

has in G a periodic solution with the primitive period ω. Then

$$\omega \geq 4/\varkappa \ . \qquad (5.3)$$

This theorem has been extended in [15].

As mentioned above, it sufficies to consider the Problem P_2 for the equation (2.18) where A_o is a bounded linear operator. Setting $h(x_o,\mu) := -A_o x_o + f_o(x_o,\mu)$ we obtain according to hypothesis (F_1) that h is lipschitzian in $\|x_o\| \leq \varepsilon$ with respect to x_o where its Lipschitz constant $l_1(\varepsilon)$ satisfies

$$l_1(\varepsilon) \leq \|A_o\| + l(\varepsilon), \qquad (5.4)$$

$$l(\varepsilon) \longrightarrow 0 \text{ as } \varepsilon \longrightarrow 0. \qquad (5.5)$$

Using (5.4) we have the following corollary.

<u>Corollary 5.2.</u> Assume the hypotheses of Problem P_2 are valid. Then any subsequence $\{\bar{\omega}_n\}$ of $\{\omega_n\}$ satisfies $\bar{\omega}_n \geq 2/\|A_o\|$ for sufficiently large n. In particular, under these conditions a subsequence $\{\bar{\omega}_n\}$ with $\lim_{n\to\infty} \bar{\omega}_n = 0$ cannot exist.

Theorem 5.1 also permit to get sufficient conditions that any subsequence $\{\tilde{\omega}_n\}$ of $\{\omega_n\}$ satisfies $\lim_{n\to\infty} \tilde{\omega}_n^{-1} = 0$. A first application yields:

<u>Corollary 5.3.</u> Suppose the hypotheses of Problem P_2 hold. Additionally we assume

(A_5). $\qquad A_o = 0 \ . \qquad (5.6)$

Then we have for any subsequence $\{\tilde{\omega}_n\}$

$$\lim_{n\to\infty} \tilde{\omega}^{-1} = 0. \qquad (5.7)$$

<u>Proof.</u> By Theorem 5.1 and according to (5.4) - (5.6) the primitive period ω_k of any periodic solution $p_k(t)$ of (2.1) sufficiently near to the origin satisfies

$$\omega_k \geq \frac{4}{l_1(\mu_k)} \ .$$

From (5.5) we get the validity of the relation (5.7).

The conditions in Corollary 5.3 imply that A_o is not invertible. There arises the question wether this property is necessary for the existence of a subsequence $\{\tilde{\omega}_n\}$ with $\lim_{n\to\infty}\tilde{\omega}_n^{-1} = 0$. To investigate this problem we assume:

(B_3). For $\mu \in (0,\mu^*)$ where μ^* is some positive number there is a family of ω_μ -periodic solutions $p_\mu(t)$ of (2.12), which depends continuously from μ and has the property

$$\|p_\mu\|_{o,c} \longrightarrow 0, \ \omega_\mu \longrightarrow \infty \quad \text{as } \mu \longrightarrow 0 . \tag{5.9}$$

(F_3).

$$\lim_{\mu \to 0} |\omega_\mu| \frac{\|f_o(p_\mu,\mu)\|_{o,c}}{\|p_\mu\|_{o,c}} = 0 \tag{5.10}$$

(A_6) All eigenvalues of A_o are semisimple.

Theorem 5.4. Asssume the hypotheses (A_1), (A_2), (A_6), (F_1) - (F_3), (S_2), (B_3) hold. Then the operator A has at least one vanishing eigenvalue.

Proof. First we proceed as in the proof of Theorem 4.2. By (A_6) $e^{A_o t}$ is uniformly bounded $\forall$ t $\in$ R. Hence, from (4.6) and (5.10) we get

$$\lim_{\mu \to 0}\left[\left\{ I - e^{-A_o\omega_\mu}\right\} \frac{q_\mu(0)}{\|q_\mu\|_{o,c}} \right] = 0. \tag{5.11}$$

By (S_2) we have

$$\lim_{\mu \to 0} \frac{q_\mu(0)}{\|q_\mu\|} = q_o$$

where

$$\|q_o\|_{o,c} = 1 . \tag{5.12}$$

Without loss of generality we can assume that A_o is in its real Jordan normal form, that is

$$A_o = \operatorname{diag}\,(A_o^0,\ A_o^1, \ldots,\ A_o^k) \tag{5.13}$$

where

$$A_o^l := \begin{pmatrix} 0 & -ib_l \\ ib_l & 0 \end{pmatrix}, \quad b_l > 0,\ l = 1,\ldots,k,$$

A_o^o is the zero-matrix provided it arises in (5.13). If we consider the relation (5.11) for each Jordan block A_o^l then there is a sequence $\{\mu_k^l\} \longrightarrow 0$ such that

$$\lim_{\mu_k \longrightarrow 0} \left\{ I - e^{-A_o \omega \mu_k} \right\}$$

exists and is invertible. Thus, we can conclude by (5.11) that the component of q_o corresponding to that block vanishes. Then it follows from (5.12) that A_o^o must occur in (5.13), that is, A_o has a vanishing eigenvalue, q.e.d.

REFERENCES

[1] Bridges. Th.J., The Hopf bifurcation with symmetry for the Navier-Stokes equations in $(L_p(\Omega))^n$, with application to plane Poiseuille flow. Arch. Rat. Mech. Anal. 106, 335 -376 (1989).

[2] Crandall, M.G., Rabinowitz, P.H., The Hopf bifurcation Theorem in infinite dimension. Arch. Rat. Mech. Anal. 67, 53 - 72 (1978).

[3] Dunford, N., Schwartz, J., Linear Operators, Part I, Interscience 1958.

[4] Friedman, A., Partial Differential Equations, Holt, Rinehart and Winston, New York 1969.

[5] Hassard, B.D., Kazarinoff, N.D., Wan, Y-H., Theory and Applications of Hopf Bifurcation. London Mathematical Society, Lect. Not. Ser. 41, 1981.

[6] Henry, D., Geometric Theory of Semilinear Parabolic Equations. Lect. Not. Math. 840, Springer Verlag , 1981.

[7] Joseph, D.D., Sattinger, D., Bifurcating time periodic solutions and their stability. Arch. Rat. Mech. Anal. 45, 79 -109 (1972).

[8] Kato, T., Perturbation Theory of Linear Operators. Berlin, Springer-Verlag, 1966.

[9] Kielhöfer, H., Hopf bifurcation at multiple eigenvalues. Arch. Rat. Mech. Anal. 69, 53 - 83 (1970).

[10] Lasota, A. Yorke, J.A., Bounds for periodic solutions of differential equations in Banach spaces., J. Diff. E qus 10, 83 - 91 (1971).

[11] Marsden, J.E., Mc Cracken, M.F., The Hopf Bifurcation and its Applications, New York, Springer-Verlag 1976.

[12] Sattinger, D., Bifurcation of periodic solutions of the Navier-Stokes equations. Arch. Rat. Mech. Anal. 41, 66 - 80 (1971).

[13] Schneider, K.R., Über eine Integralgleichungsmethode in der Theorie autonomer Schwingungen und ihre Anwendung auf Bifurkationsprobleme, Math. Nachr. 69, 192 -204 (1975).

[14] Schneider, K.R., On quasicenter manifolds of semilinear equations in Banach spaces. Math. Nachr. 122, 215-229 (1985).

[15] Schneider, K.R., A lower bound for the primitive period of self-sustained oscillations. Coll. Math. Soc. Janos Bolyai 47. Differential Equations: Qualitative Theory, 1984, 951 -970.

[16] Schneider, K.R., Wegner, B., Об асимптотическом поведении периода предельных циклов, стягивающихся в особую точку. Мат. Иссл. 36 (2), 28 - 35 (1975).

[17] Schneider, K.R., Wegner, B., Asymptotic behaviour of the period of a closed trajectory contracting to a singular point, In.: VII. Internationale Konferenz über nichtlineare Schwingungen, Band I,2. Abh. Akad. Wiss. DDR, Jahrg. 1977 Nr.4M, 215-219, Akademie-Verlag, Berlin 1977.

STABILITY OF SINGULAR DISTRIBUTED DYNAMICAL SYSTEMS WITH ANGULAR COORDINATES

V.B. Smirnova [1)]

We often come across distributed or delay dynamical systems for which the dependence between the "output" and "input" is described by means of integral or integro-differential Volterra equations. Among these a special place belongs to the so-called phase systems (or systems with angular coordinates) [1,2]. Their mathematical description can be represented as follows:

$$\dot{\sigma}(t) = \alpha(t) + \varrho\varphi(\sigma(t-s)) - \int_0^t \gamma(t-\tau)\varphi(\sigma(\tau))d\tau \quad (s \geq 0;\ t > 0), \quad (1)$$

where $\varphi(\sigma)$ is a Δ-periodic function. Thus if the function $\sigma(t)$ is a solution of equation (1) then any function $\sigma(t) + \Delta j$ (where j is an integer) is also a solution of (1).

Equations (1) are used for example to describe phase-lock loops (here $\sigma(t)$ is the phase difference of the two generators) and synchronous electric motors ($\sigma(t)$ is the phase difference of the rotating magnetic field and the rotor). Because of the specific character of phase systems multiple equilibria are made possible. Indeed, to make sure of the fact it is sufficient to suppose $\alpha(t) \to 0$ as $t \to +\infty$ and $\varphi(\sigma)$ has at least one zero in $[0,\Delta)$. So we must treat phase systems as systems with multiple equilibria.

In the case of lumped parameters a special apparatus was created [3,4] in order to solve stability, instability and oscillation problems for phase systems. The new technique involves several new methods combined with the classical direct Ljapunov method. Some elements of this technique were propagated to the case of distributed parameters [5,6], by means of a priori integral estimates method [7]. The most fruitful among the new stability methods proved to be the nonlocal reduction method [3,6]. In the paper [8] its place among other qualitative methods is shown up. Moreover, this article also contains some main results, obtained with the help of a nonlocal reduction method both for differential and integro-differentail equations. Some efficient stability theorems were proved in [3,5,9] by means of the Bakaev-Guhz procedure introduced in [10] and its modifications which appeard in [11].

The papers [5,6,9] establish sufficient conditions for the Lagrange stability and complete stability of (1) under the supposition that the

---

1) Leningrad Institute of Civil Engineers, Chair of Mathematics
198005 Leningrad, 2-ya Krasnoarmeiskaya ul. 4, USSR

nonlinear function $\varphi(\sigma)$ is at least continuous. But it is rather important for various engineering applications to study the case of a piecewise continuous nonlinear function. The investigation of this type was started in [12]. In this connection a new concept of solution has been used different from the classical one.

So let $\varphi(\sigma)$ have discontinuities of the first type at the points Δj $(j = 0, \pm 1, \pm 2, \ldots)$ and let $\varphi(\Delta j - 0) \cdot \varphi(\Delta j + 0) < 0$. We suppose that $\varphi(\sigma)$ has a continuous bounded positive derivative on the intervals $(\Delta(k-1), \Delta k)$. Let $\sup\limits_{\sigma \in (0,\Delta)} \varphi'(\sigma) = k$.

We also suppose that $\alpha(t)$ is continuous, $\varphi(t)$ is measurable and the following inclusions hold

$$\alpha(t)e^{\varkappa t} \in L_2[0,+\infty), \quad \gamma(t)e^{\varkappa t} \in L_2[0,+\infty) \qquad (\varkappa > 0). \tag{2}$$

For equation (1) an initial function is defined by

$$\sigma(t) = \sigma^0(t) \quad \text{for} \quad t \in [-h,0] \quad \text{with} \quad \sigma^0 \in C^1([-h,0)]. \tag{3}$$

In order to define the solution of (1), (3) we use here the approach which was suggested in [13]. Let us consider a set of continuously differentiable functions

$$\varphi_h(\sigma) = \frac{1}{3h} \int_{\sigma-h}^{\sigma+h} r_h(\sigma - \xi)\varphi(\xi)d\xi,$$

where

$$r_h(\xi) = \begin{cases} \frac{2}{h}\xi + 2, & -h \le \xi < -\frac{h}{2}, \\ 1, & -\frac{h}{2} \le \xi < \frac{h}{2}, \\ -\frac{2}{h}\xi + 2, & \frac{h}{2} \le \xi \le h. \end{cases} \tag{4}$$

Notice that for an arbitrary $k_0 > k$ there exists $H(k_0) > 0$ such that $\frac{d\varphi_h(\sigma)}{d\sigma} < k_0$ for all $h < H(k_0)$.

Let us consider a set of integro-differential equations

$$\dot{\sigma}_h(t) = \alpha(t) + \varrho\varphi_h(\sigma_h(t-s)) - \int_0^t \gamma(t-\tau)\varphi_h(\sigma_h(\tau))d\tau \tag{5}$$

with initial condition

$$\sigma(t) = \sigma^0(t) \quad \text{for} \quad t \in [-h,0]. \tag{6}$$

Let h tend to 0. The limit of any uniformly (with respect to $t \in [0,T']$ $(T' < T)$) convergent subsequence of $\{\sigma_h(t)\}$ let be called by us an averaged solution of (1), (3) on $[0,T)$. By virtue of the properties of $\alpha(t)$ and $\gamma(t)$ the averaged solution of (1), (3) exists and can be prolonged on any interval $[0,T)$.

In this paper the stability of the averaged solution of (1) is investigated. The results are formulated in terms of the transfer function

$$K(p) = - \varrho e^{-ph} + \int_0^\infty \gamma(t) e^{-pt}\, dt \qquad (p \in \mathbb{C}).$$

The investigation is based on the Popov a priori integral estimates method. According to its schema certain functionals, depending on $T > 0$, are constructed on the solutions of the equation. (The type of the functionals corresponds to the hypotheses about the function $K(p)$, which are frequency called inequalities.) For this set of functionals certain uniform (with respect to T) estimates are establisted and by means of the latter there are made conclusions about the solutions' behavior. In this paper we construct the functionals on the solutions of "averaged" equations (5) and demand that the estimates were uniform with respect both to T and h. In this connection the following auxiliary proposition arising from the reasoning of article [12] is of further use.

<u>Lemma 1.</u> Suppose that for all $h < H$ and all $T > 0$ the following estimate is true

$$\int_0^T [\varphi_h(\sigma_h(t))]^2\, dt < Q, \tag{7}$$

where Q depends neither on h nor on T. Then any averaged solution of (1) satisfies the relation

$$\sigma(t) \to c_j \quad \text{as} \quad t \to +\infty, \tag{8}$$

where c_j is either a zero of $\varphi(\sigma)$ or a point of discontinuity.

We shall begin with combining Popov's method and the Bakaev-Guzh procedure in order to extend the results of article [5] to the case of a discontinuous nonlinearity.

Let us denote by σ^* the zero of $\varphi(\sigma)$ on the interval $(0,\Delta)$. Let us set also

$$\Gamma = \max \{ |\int_{\sigma^*}^{\Delta} \varphi(\theta) d\theta|, \ |\int_0^{\sigma^*} \varphi(\theta) d\theta| \},$$

$$\gamma = \min \{ |\int_{\sigma^*}^{\Delta} \varphi(\theta) d\theta|, \ |\int_0^{\sigma^*} \varphi(\theta) d\theta| \}.$$

<u>Theorem 1.</u> Suppose there exist numbers $\varepsilon > 0$, $\delta > 0$ such that the inequality

$$\operatorname{Re} K(i\omega) - \varepsilon |K(i\omega)|^2 \geq \delta \qquad (i = \sqrt{-1}) \tag{9}$$

is true for all real ω and the condition

$$4\varepsilon\delta > \nu^2, \quad \text{where} \quad \nu = \frac{\Gamma - \gamma}{\Gamma + \gamma}, \tag{10}$$

is fulfilled. Then for any averaged solution of (1) the relation (8) is satisfied.

Proof. Let us repeat for every equation (5) the scheme offered in [5] for the case n = 1 and $\lambda_1 = 0$. Suppose σ_h is an arbitrary solution of the Cauchy problem (5), (6).

Introduce the notation

$$\Delta_h^+ = \{\theta : \varphi_h(\theta) \geq 0\}, \quad \Delta_h^- = \{\theta : \varphi_h(\theta) < 0\},$$

$$\Gamma_h = \max\{|\int_{\Delta_h^+} \varphi(\theta)d\theta|, \ |\int_{\Delta_h^-} \varphi(\theta)d\theta|\},$$

$$\gamma_h = \min\{|\int_{\Delta_h^+} \varphi(\theta)d\theta|, \ |\int_{\Delta_h^-} \varphi(\theta)d\theta|\},$$

$$\nu_h = \frac{\Gamma_h - \gamma_h}{\Gamma_h + \gamma_h}.$$

Inequality (9) coincides with the frequency inequality from [5], and thus for all h > 0 and all T > 0 the following estimate is true

$$I_{oh} \equiv \int_0^T \{\nu_h|\dot\sigma_h||\varphi(\sigma_h(t))| + \delta\varphi(\sigma_h(t))^2 + \varepsilon\dot\sigma_h(t)^2\} < C, \tag{11}$$

where C depends neither on T nor on h. It follows from (10) and (4) that for $h < H_o$, where H_o is small enough, there exists $\nu_o > \nu$ such that

$$2\sqrt{\varepsilon\delta} > \nu_o > \nu_h.$$

Then it follows from (10) that estimate (7) is true. Thus, by virtue of Lemma 1 Theorem 1 is proved.

By means of the Bakaev-Gruzh procedure more effective frequency inequalities than (9) can be exploited. Suppose in addition to (2) that α and γ are differentiable and

$$\dot\gamma(t)e^{\varkappa_1 t} \quad \text{and} \quad \dot\alpha(t)e^{\varkappa_1 t} \quad \text{with} \quad \varkappa_1 > 0 \quad \text{are} \quad L_2[0,\infty). \tag{12}$$

Theorem 2. Suppose (2) and (12) are true. Suppose there exist $\varepsilon > 0$, $\delta > 0$ and $\xi > 0$ such that for all real ω the following frequency inequality is true

$$\mathrm{Re}\, K(i\omega) - \xi K^*(i\omega)[K(i\omega) + k_o^{-1} i\omega] - \varepsilon|K(i\omega)|^2 \geq \delta, \tag{13}$$

where $k_o > k$ and ε and δ satisfy (10). Then the conclusion of Theorem 1 is valid.

Proof. Suppose $\sigma_h(t)$ is a solution of (5), (6). Now let us use the scheme of [5] for the case $n = 1$, $\lambda_1 = \xi$, and evaluate the sum of functionals

$$I_{oh} + I_{1h},$$

where I_{oh} has been introduced in the proof of Theorem 1 and

$$I_{1h} = \int_0^T \{\dot{\sigma}_h^2(t) - k_o^{-1} \dot{\eta}_h(t)\dot{\sigma}_h(t)\}dt, \quad \eta_h(t) = \varphi_h(\sigma_h(t)).$$

By means of the Parseval equality and of (13) it is easy to construct an estimate

$$I_{oh} + I_{1h} \leq \int_0^T R_h(t)dt,$$

where R_h is a linear combination of various products of functions α, γ, η_h, $\dot{\eta}_h$ except $\dot{\eta}_h^2$. The relations (2) and (12) give the opportunity to estimate

$$\int_0^T R_h(t)dt$$

from above by a constant which depends neither on T nor on h. Consider now the integrand of I_{1h}.

$$\dot{\sigma}_h^2 - k_o^{-1} \dot{\eta}_h\dot{\sigma}_h = \dot{\sigma}_h[\dot{\sigma}_h - k_o^{-1} \frac{d\varphi_h(\theta)}{d\theta} \dot{\sigma}_h]$$

and so for all $h < H(k_o)$

$$I_{1h} \geq 0.$$

Thus, for all $h < H(k_o)$ and all $T > 0$ the estimate (11) is true. In order to finish the proof of Theorem 2 it remains to repeat the proof of Theorem 1.

Concerning problems with a discontinuous nonlinearity, the nonlocal reduction method may be successfully applied.

With the help of this method and by means of repeating the argument of [12] for averaging (4) we can establish

Theorem 3. Let the condition (2) be fulfilled. Suppose there exist numbers $\varepsilon > 0$ and $\lambda \in (0, \pi/2)$ such that the inequality

$$\text{Re } K(i\omega - \lambda) - \varepsilon|K(i\omega - \lambda)|^2 \geq 0$$

is true for all real ω and one of the following conditions is fulfilled:

$$4\lambda\varepsilon > - (\inf_{\sigma \in (0,\Delta)} \varphi(\sigma) \;/\; (\Delta - \sigma^*); \tag{14}$$

$$2\sqrt{\lambda\varepsilon} > (\int_{\sigma^*}^{0} \varphi(\theta)d\theta)^{\frac{1}{2}} \;/\; (\Delta - \sigma^*). \tag{15}$$

Then any averaged solution of (1) is bounded on $[0,+\infty)$. If in addition there exists $\delta > 0$ such that the inequality

$$\operatorname{Re} \chi(i\omega) \geq \delta \qquad (\forall\omega \in \mathbb{R}) \tag{16}$$

is true, then the conclusion of Theorem 1 is valid.

It is possible to complicate the functionals used in [12] by means of adding to them the functionals I_{1h} used in the text of Theorem 2. It gives the opportunity to use a new frequency inequality.

Theorem 4. Let the conditions (2) and (12) be fulfilled. Suppose there exist numbers $\zeta > 0$, $k_0 > k$, $\varepsilon > 0$ and $0 < \lambda < \varkappa/2$ such that for all real ω the inequality

$$\operatorname{Re} K(i\omega - \lambda) - \xi K^*(i\omega - \lambda)[K(i\omega - \lambda) + k_0^{-1}(i\omega - \lambda)] - \varepsilon|K(i\omega - \lambda)|^2 \geq 0$$

is true and one of the conditions (14) or (15) is fulfilled. Then any averaged solution of (1) is bounded on $[0,+\infty)$. If in addition the inequality (16) is true for some $\delta > 0$, then the conslusion of Theorem 1 is valid.

References

[1] Lindsey, W.C.: Synchronization Systems in Communication and Control, Prentice-Hall, Inc., Englewood Cliffs, New Jersey 1972.

[2] Viterbi, A.J.: Principles of Coherent Communication. Mc Graw-Hill, N. Y. 1966.

[3] Gelig, A.Kh., Leonov, G.A., Jakubovich, V.A.: Stability of nonlinear systems with multiple equilibria, Moscow, Nauka 1978.

[4] Leonov, G.A., Koryakin, Ju., Reitmann, V.: Konvergenz im Mittel in Phasensystemen. Zeitschrift für Angewandte Mathematik und Mechanik №10, 1978.

[5] Leonov, G.A., Smirnova, V.B.: Asymptotic behavior of integro-differential equations with periodic nonlinearities. Sibirsk. Math. J., V. 19, №6, 1978, 1206 - 1412.

[6] Leonov, G.A., Smirnova, V.B.: The reduction method for integro-differential equations. Sibirsk. Math. J., V. 21, 1980, №4, 112 - 124.

[7] Povov, V.M.: On absolute stability of nonlinear automatic control systems. Avtomat. i telemech., V. 22, №8, 1961, 961 - 979.

[8] Leonov, G.A., Smirnova, V.B.: Nonlocal reduction method in differential equations theory. Series in Pure Mathematies. V. II. Topics in mathematical analysis. A volume dedicated to the memory of A.L. Cauchy. World Scientific. Singapore - New-York - London - Hong kong, 658 - 694.

[9] Leonov, G., Reitmann, V., Smirnova, V.: On the Global Asymptotic Stability of Ordinary and Functional Differential Equations with Multiple Equilibria. Technische Universität Dresden. Informationen. 07-01-89.

[10] Bakaev, Ju.N., Guzh, A.A.: Optimal reception of signals of frequency modulation under the Doppler effect. Radiotehnika i Elektronika, V. 10, №1, 1965.

[11] Brockett, R.W.: On the Asymptotic Properties of Solutions of Differentail Equations with Multiple Equilibria. Journ. of Diff. Equations, V. 44, 1982, 249 - 262.

[12] Leonov, G.A., Smirnova, V.B.: Certain properties of solutions of Volterra integro-differential equations with piecewise periodic nonlinear functions. In the book: Problems of contemporary theory of periodic motions. Izhevsk, 1980.

[13] Tonkova, V.S., Tonkov, E.L.: Certain properties of averaged solutions of a control system with discontinuous nonlinearity. Differenzial'nye uravnenija, V. 9, №2, 1973, 278 - 289.

Quantum Dynamics and Algebras of Unbounded Operators

W. TIMMERMANN[1)]

0. Introduction

"The past two decades have witnessed a fruitful interaction between operator algebras and mathematical physics."[12] While here operator algebras mean mainly C*- and W*-algebras, in the same period the theory of topological algebras of unbounded operators has been essentially developed This was stimulated by the algebraic approach to quantum field theory and quantum statistical physics. From the mathematical point of view algebras of unbounded operators appear in connection with representations of non-normable topological algebras, Lie algebras, and they are also related with algebras of bounded operators.
The starting point of many investigations was the paper of G.Lassner [14]. Several hundred papers published since the seventies were devoted to such aspects of the theory as: representation theory, topologization, order structure, states and state spaces, ideals, duality theory, investigation of special classes of algebras and partial algebras, commutants and so on. A survey of the present state of some areas of the theory is given in [22] which is the first book on this topic. We refer to it for further references. Despite the intensive developments there are some important tools where only sporadic and sparse results are known. Unfortunately, many of these open questions are related with fundamental objects occuring in the investigation of quantum dynamical systems (e.g. automorphisms, groups of automorphisms, derivations, KMS-states etc.).
The aim of this contribution is to report on some known results in these fields and indicate some questions which are to be clarified. It should be emphasized that this is a very personal choice of the material and the representation is limited by place as well as by the knowledge of the author.

1. Preliminaries

For a dense linear manifold D in a (separable) Hilbert space H the set of linear operators $L^+(D) = \{a: aD \subset D, a^*D \subset D\}$ forms a *-algebra with respect to the usual operations and the involution $a \to a^+ = a^* \restriction D$.
An Op*-algebra A(D) is a *-subalgebra of $L^+(D)$ containing the identity operator 1. The graph topology t_A on D induced by A(D) is given by the seminorms

$$D \ni \varphi \longrightarrow \|a\varphi\| \quad \text{for all } a \in A(D).$$

1) Technical University of Dresden, Department of Mathematics, Mommsenstr. 13, Dresden, DDR-8036

For $A(D) = L^+(D)$ write t instead of t_A.
On Op*-algebras many topologies can be defined. The philosophy can be expressed as follows: any range of problems requires its own specific topology.
We mention only topologies needed in the sequel. Consider the following seminorms

$$a \to \|a\|_M = \sup_{\varphi,\psi \in M} \langle \varphi, a\psi \rangle$$

If M runs over all t-bounded (relatively t-compact) subsets of D we obtain the topology τ_D (τ_D^c) in $L^+(D)$.
An Op*-algebra A(D) is said to be

<u>closed</u> if $D = \bigcap_{a \in A} D(\bar{a})$, $\bar{a}$ denotes the closure of a,

<u>selfadjoint</u> if $D = \bigcap_{a \in A} D(a^*)$

For an arbitrary *-algebra R the set

$$P(R) = \{\text{finite sums of elements } a^*a,\ a \in R\}$$

denotes the (natural) positive cone in R.
For Op*-algebras A(D) we have also the cone

$$K(A(D)) = \{a \in A(D): \ \langle a\varphi, \varphi \rangle \geq 0 \text{ for all } \varphi \in D\}$$

We use the notions P(A),(P(D),K(A),K(D), resp.) instead of P(A(D)), $(P(L^+(D),\ldots,K(L^+(D)$, resp.).
Clearly, $P(A) \subset K(A)$ and in contrast to the C*-case the inclusion may be proper.
By $M_n(R)$ we mean the *-algebra of all (nxn)-matrices with elements from R and $P_n(R)$ denotes the corresponding cone ($K_n(D)$ means the corresponding cone in case $R = L^+(D)$).
By a quantum dynamical system we mean a triple (R, α, G) where $R[\tau]$ is a topological *-algebra, G a (topological) group, $\alpha : G \to \mathrm{Aut}(R)$ a homomorphism of G into the group of automorphisms of R (and in addition there should be required some continuity).
We will use only one-parameter groups of automorphisms, i.e. $G = \mathbb{R}$ (call these shortly automorphism groups and write (R, α) in this case).
A *-automorphisms of R is an automorphism α with $\alpha(a^+) = \alpha(a)^+$.
There are several possible notions of continuity of automorphism groups (cf. [2]). For simplicity we use the following definition:
A continuous one-parameter group (α_t) of *- automorphisms on a topological *-algebra $R[\tau]$ is a family (α_t), $t \in \mathbb{R}$, of continuous automorphisms of R with:
$\alpha_0 = \mathrm{id}$ (identical automorphism), $\alpha_{s+t} = \alpha_s \circ \alpha_t$ for all $s, t \in \mathbb{R}$,
$\lim_{t \to 0} \alpha_t(a) = a$ for all $a \in R$ (in the C*-case this is a strongly continuous automorphism group).
In the next sections we will discuss automorphisms and automorphism groups derivations, positive and completely positive maps as well as dissipations.

2. Automorphisms, automorphism groups and derivations

In what follows we consider only *-automorphisms on Op*-algebras.
There are several notions which express some regularity of an automorphism α on A(D): α is said to be spatial (inner), if $\alpha(a) = u^{+}au$ for all $a \in A(D)$ and some $u \in L^{+}(D)$ ($u \in A(D)$).
It is easy to see that $\alpha(P(A)) \subset P(A)$ and if α is spatial then $\alpha(K(A)) \subset \subset K(A)$. The only general result concerning the implementability of automorphisms is the following one ([30],[22]).

Proposition 1

Let A(D) be a closed Op*-algebra with cyclic vector $\varphi_0 \in D$ (i.e. $\{A\varphi_0\}$ is t_A-dense in D). Let f_{φ_0} be the induced vector state on A(D), α - a *-automorphism on A(D). If there is a vector $\psi_0 \in D$ such that $f_{\varphi_0} \circ \alpha = f_{\psi_0}$, then:

i) there is an isometry $u \in L^{+}(D)$ with $\alpha(a) = u^{+}au$ for all $a \in A$

ii) if A(D) is selfadjoint, then in additional $(\bar{u}u^{*})a\varphi = a(\bar{u}u^{*})\varphi \quad \forall \varphi \in D$

iii) if A(D) is selfadjoint and has trivial commutant then u in i) is unitary.

For a discussion of the many possible notions of commutants in the context of algebras of unbounded operators we refer to the literature.
The assertion of the proposition holds-roughly speaking-if A contains a sufficient number of operators (e.g. all finite dimensional operators on D)
As a corollary it follows that any *-automorphism on $L^{+}(D)$ is inner, what has been well-known for a long time [35].
While in the bounded case *-automorphisms are norm-continuous, there are non-continuous automorphisms (with respect to τ_D or the weak or the strong topology) in the unbounded case. Spatial (inner) automorphisms are continuous with respect to many topologies and this is in general easy to see.
Only for the algebra of polynomials (appropriately realized as operator algebra) and the Weyl algebra there are several non-trivial results concerning the implementability and continuity (cf. e.g. [2], [7], [25], [27]).
Next let us discuss some problems related with groups of *-automorphisms. Here the situation is also not very hopeful. In the mathematical formulation of quantum mechanics time evolution of observables is given by

$$a \longrightarrow \alpha_t(a) = e^{ith}ae^{-ith}$$

where h is the (selfadjoint) Hamiltonian of the system under consideration
Clearly, for B(H) no problems arise to define this automorphism group.
But the situation dramatically changes if we turn to algebras of unbounded operators. Several aspects of this problem were discussed in [8], [9], [25], [26] and recently in [2].

Consider only $L^+(D)$. To identify α_t as automorphisms we must guarantee that D is invariant under e^{ith}. This is very hard. The results in the papers just mentioned are proved for domains D of the following kind:

$$D = \bigcap_{n \geq 0} D(t^n) \ , \quad t = t^* \geq 1$$

The first step is to find conditions which imply $e^{ith} H_k \subset H_k$, where (H_k) is the Hilbert scale induced by t, i.e. $H_k = D(t^k)$ equipped with the scalar product $\langle \varphi , \psi \rangle_k = \langle t^k \varphi , t^k \psi \rangle$.
Such conditions are e.g.:
i) $D \subset D(h)$ and there is a sequence (c_k) of positive numbers with

$$\pm i[h, t^k] \leq c_k \ t^k \text{ for all } k$$

ii) as i) but the estimation is replaced by

$$\| [t, k]_k \varphi \| \leq c_k \ \| t^k \varphi \| \quad \text{for all } \varphi \in D, k \in \mathbb{N}.$$

Here $[\ ,\ ]_k$ stands for the k-fold commutator.
The next step is the following one. Suppose $e^{ith} \in L^+(D)$. How can h be pertubed (i.e. $h \rightarrow h+v$) such that $e^{it(h+v)} \in L^+(D)$, too? This problem is also discussed in [25], mainly for concrete examples.

Now we turn to generators of automorphism groups. We recall a definition: A linear operator δ from a $*$-subalgebra $D(\delta) \subset R$ in R is called a derivation ($*$-derivation) if $\delta(xy) = \delta(x)y + x\,\delta(y)$ for all $x, y \in R$ (and $\delta(x^*) = \delta(x)^*$).
There is a general result which can be proved straightforward:

Proposition 2
Let $R[\tau]$ be a topological ($*$-)algebra and (α_t) a continuous group of ($*$-)automorphisms on $R[\tau]$. Define a linear operator δ on $D(\delta)$ by

$$D(\delta) = \{ x \in R : \lim_{t \to 0} (1/t)(\alpha_t(x) - x) \text{ exists}$$

$$\delta(x) = \lim_{t \to 0} (1/t)(\alpha_t(x) - x) \quad \text{for all } x \in D(\delta)$$

Then δ is a ($*$-) derivation and $D(\delta)$ is a ($*$-)subalgebra of R.
But to prove that $D(\delta)$ is dense in $R[\tau]$ it must be supposed that $R[\tau]$ is sequentially complete. Thus, we see that for quantum dynamical systems the infinitesimal generators are $*$-derivations.
The theory of derivations in C^*- and W^*-algebras has been developed into two directions. First there where considered derivations **on** such algebras (for some important results cf. [21]). But soon it became clear that unbounded derivations (defined on dense subalgebras of C^*- or W^*-algebras) are appropriate objects also with regard to applications in quantum statistical physics. Thus, in the past 10-15 years a rich theory was elaborated.

On the other hand there are only sporadic results concerning derivations on non-normable topological algebras. Let us a little bit comment on this situation. The first result on derivations on a rather restrictive class of Op*-algebras was obtained in [5]. In [10] there were considered derivations on so-called EC*-algebras (roughly: algebras in which the bounded elements form a C*-algebra) and [11] was devoted to unbounded derivations in C*- and W*-algebras which can be extended to derivations on these algebras but with ranges in certain spaces/algebras of unbounded operators.

The only general result is the following one [10],[21] (cf. also Proposition 1):

Proposition 3

Let A(D) be an Op*-algebra which contains all finite dimensional operators from $L^+(D)$. Then any derivation δ on A(D) is spatial, i.e. $\delta(a) = [a,h]$ for some $h \in L^+(D)$.

If δ is a *-derivation then $h = h^+$.

Hence, any derivation on $L^+(D)$ is inner [13],[35].

In [2] Becker investigated derivation on a number of concrete algebras, e.g. $C^\infty([0,1])$, $C_0^\infty([0,1])$, $C^\infty(\mathbb{R})$, $C_0^\infty(\mathbb{R})$ and on the corresponding matrix algebras $M_n(A)$ (where A runs over the algebras just mentioned). For such derivations explicit descriptions can be given.

Moreover, on LMC*-algebras any derivation is continuous and generates a one-parameter group of automorphisms. The question under which conditions a derivation on an Op*-algebra will be a generator of an automorphism group is very difficult and completely open (exept trivial cases New ideas are necessary to push forward the investigation of derivations on topological algebras.

Let us close this section with the remark that derivations are related with another interesting topic, namely commutators. This is clear because inner or spatial derivations are represented as commutators, namely $\delta(a) = [a,h]$. In the bounded case there are many important results on commutators. The study of commutators in algebras of unbounded operators was started with [31], and results obtained so far are published in [23],[24]. Finally, there are some interesting questions concerning the range of a derivation $\delta = [.,h]$. Up to now this problem has not been considered in algebras of unbounded operators.

3. Positive and completely positive maps

It is well-known that positivity is an important feature in the theory of operator algebras.

Let us recall some facts needed in the sequel.

A linear map F from a $*$-algebra R in $L^+(D)$ is called

- $(P(R),P(D))$-positive ($(P(R),K(D))$-positive, resp.) if $F(x) \in P(D)$ for all $x \in P(R)$ ($F(x) \in K(D)$ for all $x \in P(R)$ resp.)
- $(P(R),P(D))$-n-positive if $F(x_{ij}) \in P_n(D)$ for all $(x_{ij}) \in P_n(R)$ (i.e. $F_n = F \otimes id$ is a $(P_n(R),P_n(D))$-positive map from $M_n(R)$ into $M_n(L^+(D))$).
- $(P(R),K(D))$-n-positive if for all $(x_{ij}) \in P_n(R)$

$$\sum_{i,j=1}^{n} \langle F(x_{ij})\varphi_i, \varphi_j \rangle \geq 0 \quad \text{for all } \varphi_1, \ldots, \varphi_n \in D.$$

Finally, F is said to be $(.,.)$-completely positive if it is $(.,.)$-n-positive for all n. (here $(.,.)$ stands for any possible pairing of the cones above). Remark that for τ_D-continuous maps some of the notions of positivity coincide (cf. e.g. [32]). Use c.p. for "completely positive".
The structure of single positive maps as well as of the set of all positive maps is very complicated and not clarified already in the case of maps between low dimensional matrix algebras (cf. e.g. [34], [36]).
By means of contrast the structure of completely positive maps between C -algebras was cleared up by the famous Stinespring theorem [28].
This theorem was extended to the unbounded case by Lassner/Lassner [15] and Powers [20]. To avoid additional notion and notations we recall here the variant of [15] only for $L^+(D)$.

<u>Proposition 1</u>

Let F be a $(P(R),K(D))$-completely positive map from a $*$-algebra R into $L^+(D)$. Then there exist a closed $*$-respresentation π from R into $L^+(D_1)$ and a continuous map V from D[t] into $D_1[t]$ such that

$$\langle \varphi, F(x)\psi \rangle = \langle V\varphi, \pi(x)V\psi \rangle \text{ for all } x \in R,\ \varphi, \psi \in D.$$

Moreover $\{\pi(R)VD\}$ is t-dense in D .

Let us remark that the rôle of c.p. maps for physics was discussed by several authors (cf.e.g. [16]).
In [1] Arveson studied among other things the order structure of the set of c.p. maps between C*-algebras. There are several extremal problems associated with c.p. maps. Note three of them:

i) extreme rays of the cone of c.p. maps
ii) extreme points of the order interval $[0,F]$ for a fixed c.p. map F
iii) extreme points of the set of c.p.maps which map $e \in R$ into a fixed operator B.

Some generalizations to the unbounded case were considered in [32].
Let us mention two further results which were obtained by E.Scholz and will be published in Mathematische Nachrichten in the near future.

Proposition 2

Let K be a finite dimensional Hilbert space and F a τ_D-continuous, c.p. map from $L^+(D)$ into $B(K)$. Then there exists a unique decomposition $F = F_1 + F_2$ such that

i) F_1, F_2 are c.p.

ii) F_1 is τ_D^c-continuous

iii) F_2 is τ_D-continuous and singular.

F_2 singular means that $F(c) = 0$ for all $c \in C(D)$. Here $C(D)$ is the τ_D-closure of the *-ideal of finite dimensional operators from $L^+(D)$.
It is an open question what will happen if we drop the condition that K is finite dimensional or if we consider c.p. maps between Op*-algebras.
To describe the next result we need some further notions. A c.p. map F from $L^+(D)$ into $B(K)$, $\dim K = n$, $F(1) = 1$ is called an n-vector state if there is a linear map v from K to D so that

$$\langle \varphi, F(a)\psi \rangle = \langle v\varphi, av\psi \rangle \text{ for all } a \in L^+(D), \quad \varphi,\psi \in K.$$

Denote the set of all n-vector states by V^n. By τ we denote the so-called BW-topology on the set of all τ_D-continuous linear maps from $L^+(D)$ to $B(K)$ (cf. [1] for the bounded case). This topology can be given by the following basis of neighbourhoods of zero:

$U(m, \varepsilon, a_j) = \{G: L^+(D) \to B(K) \text{ with } \| G(a_j) \| < \varepsilon, \ 1 \leq j \leq m\}$

Then the following proposition is valid.

Proposition 3

Let $\bar{V}^n$ be the τ-closure of V^n. Then $\bar{V}^n$ is the set of all F with $F = F_1 + F_2$, $F_1 : L^+(D) \to B(K)$, c.p.;

$F_1(I) + F_2(I) = I$;

$F_1(a) = w^*aw$, where $w: K \to D$ is a contraction;

F_2 is singular and τ_D-continuous.

The corresponding result for states (i.e. n = 1) was proved in [18].

4. Dissipations

We close with a short section on dissipations. Because we are interested in applications to topological *-algebras we do not discuss dissipativity in the context of (nonlinear) classical dynamical systems. This interesting feature is considered in other contributions in this volume.
There are three closely related notions/concepts:

- dissipative operators (on Banach spaces)
- dispersive operators (on real ordered Banach spaces)
- dissipations (on C^*-algebras).

Let us mention that the concept of locality provides a unification of these seemingly disparate mathematical descriptions of the physical notion "dissipation" (cf.e.g.[4]).
Dissipations are generalizations of derivations, display many properties of elliptic differential operators and are candidates for generators of contractive semigroups. A comprehensive review of the state of the theory of dissipations in C*-algebras is given in [3].
Up to now dissipations in algebras of unbounded operators were not considered. It is our intention to start such investigations in a forthcoming paper. As a first step we give some definitions and results which are of more or less pure algebraic character.
Let A be a *-algebra, Δ a linear operator from a *-subalgebra $D(\Delta)\subset A$ into A so that

$$\Delta(a^*) = \Delta(a)^*$$
$$\Delta(a^*a) \leq \Delta(a)^*a + a^*\Delta(a) \quad \forall a \in D(\Delta) \qquad (1)$$

Then Δ is called a dissipation.
In general one would suppose that A is a topological *-algebra and $D(\Delta)$ is dense in A. Remark that some authors use "$\geq$" in (1).
Clearly, *-derivations are special cases of dissipations and Δ is a *-derivation if and only if Δ is conservative, i.e. $\pm\Delta$ are dissipative.
The following fact is straightforward to verify.

<u>Lemma 1</u>
Let δ be a *-derivation on A so that $D(\delta^2)$ is a *-subalgebra . Then

$$\Delta = \lambda_0 + \lambda_1\delta + \lambda_2\delta^2 \qquad (2)$$

is a dissipation if $\lambda_0 \geq 0$, $\lambda_2 \leq 0$.

This should be compared with the ellipticity condition if one put on appropriate function algebras $\delta = d/dt$.
Remark that even for C*-algebras it is nontrivial to decide under which conditions a dissipation has form (2) (cf.e.g. [4])
In the sequel we consider only dissipations defined on the whole algebra A. Then the set of all dissipations forms a convex cone. Analogously to completely positive maps we introduce complete dissipations. Let be a dissipation on A, put $\Delta_n = \Delta \otimes id_n$ on $A \otimes M_n = M_n(A)$.
Then Δ is called a complete dissipation (completely dissipative) if is dissipative for all n.
Lindblad ([16],[17]) introduced several dissipation functions. For a given dissipation Δ put

$d(\Delta)$: $d(\Delta;x,y) = \Delta(x^*)y + x^*\Delta(y) - \Delta(x^*y) \quad \forall\ x,y \in A.$

Then:

Δ is a dissipation if and only if $d(\Delta;x,x) \geq 0$ (for all $x \in A$)
Δ is a derivation if and only if $d(\Delta;x,x) = 0$ (for all $x \in A$)
Δ is a complete dissipation if and only if $d(\Delta_n,x,x) \geq 0$ for all n and all $x \in M_n(A)$.

(Remark that Lindblad includes $\Delta(e) = 0$ in the definition of complete dissipations).

We add some propositions which are analogous to the bounded case ([16], [17]). The first one describes to which extent the dissipation function d determines .

Proposition 2

Suppose A is a *-algebra so that any derivation on A is inner or A(D) is an Op*-algebra so that any derivation on A(D) is spatial. If Δ, Δ' are dissipations with $d(\Delta) = d(\Delta')$ then $\Delta = \Delta' + [.,y]$ where $y = ih$, $h = h^+$, $h \in A$ or $h \in L^+(D)$.

The proof is based on the observation that $d(\Delta) = d(\Delta')$ implies $\Delta - \Delta'$ is a derivation.

The following proposition provides the construction of a large set of complete dissipations.

Proposition 3

Let F be a completely postitive map on $L^+(D)$ (resp. on A(D) so that for F the Stinespring representation holds, cf.sect.3). Then for any $h = h^+$ $\in L^+(D)$ ($\in A(D)$, resp.)

$$\Delta(x) = (1/2)\{F(1), x\} - F(x) - i[h, x]$$

is a complete dissipation.

Finally we add a result concerning the representation of a further type of dissipation functions.

Proposition 4

Let A(D) be an Op*-algebra and d': $A \times A \times A \rightarrow L^+(D)$ a trilinear map satisfying the following complete positivity condition:

$$\sum_{i,j=1}^{n} z_i^+ d'(y_i^+, x_i^+ x_j, y_j) z_j \geq 0 \qquad (3)$$

for all n-tuples $(x_i), (y_i), (z_i) \subset A(D)$, $\forall n$.

Then there exist a *-representation π of A(D) in $L^+(D_1)$ and a linear map V: $A(D) \rightarrow L(D, D_1)$ so that

$\langle d'(x,y,z)\varphi, \psi\rangle = (\pi(y)V(z)\varphi, V(x^+)\psi)$

for all $\varphi, \psi \in D$. Here (,) denotes the scalar product in the Hilbert space $D_1 = K$. Moreover, V is continuous from D[t] into $D_1[t_B)$, $B = \pi(A)$.

The proof is similar to that of the Stinespring theorem.

We close the section with the remark that the dissipation funtions d, d′ are closely related with cohomology of operator algebras.

References

[1] Arveson, W.: Subalgebras of C*-algebras. Acta Math. 123(1969), 141-244.

[2] Becker, R. : Derivationen und KMS-Zustände auf unbeschränkten Operatorenalgebren. Dissertation A, Leipzig 1989.

[3] Bratteli, O.:Derivations, Dissipations and Group Actions on C*-Algebras. Lecture Notes in Mathematics 1229, Berlin 1986.

[4] Bratteli, O.,G.A.Elliott, D.W.Robinson: The characterization of differential operators by locality; dissipations and ellipticity. Preprint, Trondheim 1985.

[5] Brödel, Chr., G.Lassner: Derivationen auf gewissen Op*-Algebren. Math.Nachr. 67 (1975), 53-58.

[6] Choi,M.D.: Positive linear maps. in: Operator Algebras and Applications.ed.:R.Kadison, American Math.Soc. 1982.

[7] Dixmier,J.: Sur les algebres de Weyl. Bull.soc.math.France 96 (1968), 209-242.

[8] Faris,W.G., R.B.Lavine: Commutators and selfadjointness of Hamilton operators. Comm.Math.Phys. 35 (1974), 39-48.

[9] Fröhlich,J.: Application of commutator theorems to the integration of representations of Lie algebras and commuation relations. Comm.Math. Phys. 54 (1977), 135-150.

[10] Inoue,A., S.Ota: Derivations on algebras of unbounded operators. Trans.Amer.Math.Soc. 261 (1980), 567-577.

[11] Inoue,A., S.Ota, J.Tomiyama : Derivations of operator algebras into spaces of unbounded operators. Pac.J.Math. 96 (1981), 389-404.

[12] eds.: Joergensen,P.E.T.,P.S.Muhly: Operator Algebras and Mathematical Physics. Providence 1987.

[13] Kröger,P.: Derivationen in $L^+(D)$. Wiss.Z.KMU,Math.-Nat.R.24 (1975),525-528.

[14] Lassner,G.: Topological algebras of operators. Rep.Math.Phys. 3 (1972), 279-293.

[15] Lassner,G.,G.A.Lassner:Completely positive maps and unbounded observables. Rep.Math.Phys. 11 (1977), 133-140.

[16] Linblad,G.: On the generators of quantum dynamical semigroups. Comm. Math.Phys. 48 (1976), 119-136.

[17] Lindblad,G.: Dissipative operators and cohomology of operator algebras. Lett.Math.Phys. 1 (1975/76), 219-224.

[18] Löffler,F.,W.Timmermann: On the structure of the state space of maximal Op*-algebras. Publ.RIMS, 22 (1986), 1063-1078.

[19] Pedersen,G.K.: C*-Algebras and Their Automorphism Groups. London-New York 1979.

[20] Powers,R.T.: Selfadjoint algebras of unbounded operators. I. Comm. Math.Phys. 21 (1971), 85-124. II. Trans.Amer.Math.Soc. 187 (1974), 261-294.

[21] Sakai, S.: C*-Algebras and W*-Algebras. Berlin 1971.
[22] Schmüdgen, K.: Unbounded Operator Algebras and Representation Theory. Berlin 1989.
[23] Scholz, E., W. Timmermann: Additive and multiplicative commutators in algebras of unbounded operators. Math. Nachr. to appear.
[24] Scholz, E., W. Timmermann : Special additive and multiplicative commutator representations for diagonal operators. Math. Nachr. to appear.
[25] Schröder, M.: Automorphismen von Algebren unbeschränkter Operatoren. Diss. A, Leipzig 1981.
[26] Schröder, M., W. Timmermann: Invariance of domains and automorphisms in algebras of unbounded operators.
in: Proc. Int. Conf. on Operator Algebras and Group Representations, Boston-London-Melbourne 1984
[27] Schröder, M., W. Timmermann: Einige einfache Eigenschaften von Automorphismen in Algebren unbeschränkter Operatoren. Wiss. Z. KMU, Math.-Nat. R. 31 (1982), 69-73.
[28] Stinespring, W. F.: Positive functions on C*-algebras. Proc. Amer. Math. Soc. 6 (1955), 211-216.
[29] Stormer, E.: Positive linear maps of operator algebras. Acta Math. 110 (1963), 233-278.
[30] Takesue, K.: Spatial theory for algebras of unbounded operators. Rep. Math. Phys. 21 (1985), 347-355.
[31] Timmermann, W.: On commutators in algebras of unbounded operators. Z. Analysis u. Anw., 7 (1988), 1-14.
[32] Timmermann, W.: On positive maps in algebras of unbounded operators. Math. Nachr., to appear.
[33] Timmermann, W.: On Cauchy-Schwarz inequalities for positive maps in algebras of unbounded operators. JINR Communications, E5-87-294, Dubna 1987.
[34] Tomiyama, J.: On the geometry of positive maps in matrix algebras. in: [11], 357-364.
[35] Uhlmann, A.: Properties of algebras $L^+(D)$. JINR Communications E2-8149, Dubna 1974.
[36] Woronowicz, S. L.: Positive maps of low dimensional matrix algebras. Rep. Math. Phys. 10 (1976), 165-183.

Mechanics from a system theoretic point of view

G. Wunsch[1)]

1. Introduction

To provide some unity in the exciting but very complex field of time-varying processes, observed in nature and engineering, we treat in this paper such processes as special cases of a rather general class of processes, which, for system theoretic reasons, we call Markov processes. Throughout this paper, we regard Markov processes as a mathematical structure which can be characterized by a partial two-parametric semigroup. It is our claim that many time-varying processes, observed in nature and engeneering, can be considered as special cases of Markov processes just by imposing additional assumptions on the set theoretic and algebraic properties of their partitial two-parametric semigroups.

So far most of the publications, devoted to the theory of processes, are concerned with processes observed in engineering. Because of this, the concept of Markov processes has been applied to such processes nearly exclusively. To overcome this situation, we relate in this paper the concept ofMarkov processes to classical and non-classical mechanics. It turns out that this approach allows for a simple and novel outlook on processes observed in mechanics. For that reason, we believe that the theory of Markov processes can be considered as a metatheory, which helps to provide some unity in the fundamental laws of nature and engineering.

Before proceeding to technical definitions, we like to stress the importance of invariant functionals to the description of highly complex Markov processes. Among these, integral invariances are of special significance and consequently, special emphasis is on them in this paper.

2. Definition of processes

Definition 1. Let $T = (T, \leq)$ be a lineary ordered set, $T' \subset T$ an intervall of T, and X an arbitrary set. Then any set Ξ of signals $\xi : T' \to X$, $\xi(t) = x$ is called a process over the time set T taking its values x in the phase space X.

As a consequence of Definition 1, in general, we have

$$\Xi \subset \bigcup_{T' \subset T} X^{T'} := \langle T, X \rangle \tag{1}$$

or, in special cases, $\Xi \subset X^T$. If a signal ξ is an element of a

1) Technische Universität Dresden, DDR-8027, Mommsenstraße 13

given process Ξ, then we call ξ a realization of the process Ξ.

A typical example of a process with T = X = $\mathbb{R}$ is provided by the set of all solutions to an ordinary differential equation $\dot{\xi}(t) = f(t, \xi(t))$.

The classification and description of processes can be done in many different ways. However, such a classification should yield a clear formalization of many, more or less vague, notions used in physics and philosophy, like determinism and reversibility /2/,/5/. Because we may relate to any process Ξ a family α of phase relations

$$A_t := \{ x \mid \bigvee_{\xi \in \Xi} x = \pi_t(\xi) \} \subset X^n \quad , x \in X^n, t \in T^n, \; n \in \mathbb{N} \tag{2a}$$

with

$$\pi_t(\xi) := (\xi(t_1), \dots, \xi(t_n)) \tag{2b}$$

in an unique way, any classification should take the properties of this family into consideration.

The notion of a process, as just defined, is very general. To get good results and interesting concepts, we must particularize and impose special structure. As already mentioned in the introduction this is done in this paper by a detailed processes are completely characterized by their two-parametric phase relation $A_t \subset X^2$, $t = (t_1, t_2)$ and consequently, we derive the following definitions.

Definition 2. An universal algebra (T, X, α)

$$\alpha : T_+^2 \rightarrow \mathcal{P}(X^2) \quad , \alpha(t_1, t_2) = A_{t_1, t_2} \tag{3}$$

is called a Markov semigroup, if we have for all $t_1, t_2, t_3 \in T$ with $t_1 \leq t_2 \leq t_3$

$$A_{t_1, t_3} = A_{t_1, t_2} \circ A_{t_2, t_3} \tag{4}$$

where $\circ$ denotes the usual composition of relations.

According to this definition, for Markov semigroups, the mapping α in (3) is a special mapping of $T_+^2 \subset T^2$ into the semigroup $(\mathcal{P}(X^2), \circ)$. Consequently, the mapping α can also be considered as a two-parametric family of relations $(A_{t_1, t_2})_{t_1, t_2}$ $t_1, t_2 \in T^2$, which constitutes a partially defined semigroup with respect to the composition $\circ$.

Thus (4) implies

$$A_{t,t} \subset I \quad , N(A_{t_1, t_2}) \subset V(A_{t_2, t_2}) \tag{5}$$

where I denotes the identity relation on X and N(.), V(.) the image, domain of (.) respectively.

Definition 3. A process $\Xi \subset \langle T, X \rangle$ is called a Markov process, if there is a Markov semigroup (T,X,α) such that

$$\xi \in \Xi :\Leftrightarrow \bigwedge_{t_1,t_2 \in T'} (\xi(t_1),\xi(t_2)) \in A_{t_1,t_2} \quad , T' = D(\xi) \tag{6}$$

where T' denotes the domain $D(\xi)$ of ξ.

Note that any Markov process Ξ can be defined directly in terms of a formal language by the properties of its realizations /2/.

In many applications, the relations $A_{t_1,t_2} \subset X^2$ of α are functional. In this case, we call the corresponding process a deterministic Markov process and rewrite the phase relation A_{t_1,t_2} as a mapping F, i.e.

$$x_2 = F(t_2,t_1,x_1) :\Leftrightarrow (x_1,x_2) \in A_{t_1,t_2}. \tag{7}$$

With $A_{t_1,t_2} := F(t_2,t_1,.)$, (3) and (4) can be reformulated analogously. Clearly, F is a mapping of the form

$$F : U \subset T_+^2 \times X \to X , \qquad U = D(F) \tag{8a}$$

and maps any value x_1 of a realization at t_1 into a new value x_2 at $t_2 > t_1$. A physical interpretation of this fact yields that the future of a realization ξ depends exclusively on the present value $\xi(t_1)$ of ξ and not on the past of ξ. Because of this, the mapping F is called the transition operator of the deterministic Markov process.

In view of the above discussion, we symbolize now any deterministic Markov process with (T,X,F). Additionally, we like to mention that, by (5), we get

$$F(t,t,x) = x \tag{8b}$$

for all $x \in D(F(t,t,.))$ and for all $t \in T$.

3. Classification

We shall find it convenient to classify deterministic Markov processes (T,X,F) according to the properties of the sets T and X, and according to the properties of the mapping F.

For example, for digital systems, one assumes in general $T = \mathbb{Z}$ and sometimes even X = finite set. In physics and especially in classical mechanics, the sets T and X are topological spaces, i.e. $T = \mathbb{R}$, $X = \mathbb{R}^n$, and the mapping F is continuous with respect to the appropriate topologies.

In many applications of analog systems and deterministic physics, the relations $A_{t_1,t_2} = F(t_2,t_1,.)$ are even injective. Under this condition, the corresponding deterministic Markov process is called reversible. Clearly, in (3), the image $(\mathcal{P}(X^2), \circ)$ of the mapping α is now given by the semigroup $\bigcup_{B \subset X} (X^B, \circ)$, which contains with $f: B \to X$ the inverse mapping f^{-1}.

Consequently, the values $\alpha(t_1,t_2) = F(t_2,t_1,.)$ of the mapping α satisfy

$$F^{-1}(t_2,t_1,\cdot)\circ F(t_2,t_1,\cdot) = F(t_1,t_1,\cdot). \tag{9a}$$

If we set formally

$$F^{-1}(t_2,t_1,\cdot) = F(t_1,t_2,\cdot) \tag{9b}$$

(4) becomes with $\Lambda_{t_1,t_2} = F(t_2,t_1,.)$ for all $(t_1,t_2)\in T^2$ (9c)

and especially

$$F(t_1,t_2,\cdot)\circ F(t_2,t_1,\cdot) = F(t_1,t_1,\cdot) \tag{9d}$$

4. Classical mechanics

Throughout this section, we consider Markov semigroups of the type $(T,X,F) = (\mathbb{R}, \mathbb{R}^n, F)$ where F belongs to a deterministic and reversible Markov process Ξ such that the equations (9) and (8b) are satisfied.

If we restrict our attention to smooth processes, that is to processes where $F\in C^r$ $(r>1)$ and $D(F) = U$ is an open set in $T^2\times\mathbb{R}^n$, then such a process Ξ can be characterized by the vector field

$$g(t,x) = \frac{\partial}{\partial t_2} F(t_2,t_1,x)\Big|_{t_2=t_1}. \tag{10a}$$

Thus, any realization ξ satisfies /6/

$$\dot{\xi}(t) = g(t,\xi(t)) \qquad (t,\xi(t))\in U. \tag{10b}$$

Now, let us consider the realization ξ as a trajectory in the phase space of classical mechanics, i.e. $\xi(t) = (q(t),p(t))$. This implies that the Hamiltonian vector fields of classical mechanics are just special case of (10a), and, consequently, the dynamic of classical mechanics can be described by a special deterministic, reversible, and smooth process.

Theorem 1. Let $T = \mathbb{R}$ and let Ξ taking its values in $X = \mathbb{R}^n$. Then Ξ can be regarded as a process in the sense of classical mechanics, if its realizations are solutions of (10b) subject to the restrictions:

a) n is even: $n = 2m$, $m\in\mathbb{N}$

b) the vector field $g: U\subset T\times\mathbb{R}^n\to\mathbb{R}^n$ is the Hamiltonian field of a differentiable scalar field $H: U\subset T\times\mathbb{R}^n\to\mathbb{R}$, i.e.:

$$g(t,x) = \mathrm{grad}^* H(t,x) \tag{11a}$$

with

$$\mathrm{grad}^* = \left(\frac{\partial}{\partial x_{m+1}},\dots,\frac{\partial}{\partial x_{2m}}, -\frac{\partial}{\partial x_1},\dots,-\frac{\partial}{\partial x_m}\right). \tag{11b}$$

The $x_i = p_i$'s are velocities, $x_{m+i} = q_i$.

5. Statical Mechanics

Again, let $(T,X,F) = (\mathbb{R}, \mathbb{R}^n, F)$ where we assume in this paragraph that F is continuous from the right, and that the limit on the right exists, i.e.

$$\lim_{t_2 \to t_1+} \frac{F(t_2,t_1,x)-x}{t_2-t_1} = g(t_1,x). \quad (12)$$

Additionally, the vector field g should statisfy $g(t,.) \in C^r$ $(r>1)$.

In the sequel, we call such a Markov process a continuous, deterministic Markov process. Clearly, a continuous, deterministic Markov process is in general not reversible and has noncontinuous realizations.

Let $\varphi : T \times \mathbb{R}^n \to \mathbb{R}$ be a smooth functional on a continuous Markov process $(\mathbb{R}, \mathbb{R}^n, F)$. Then, according to /7/, the following generalization of Liouville's theorem can be proved:

For all $V_t \subset U_t$ (U_t = image of U by t), we have

$$I_\varphi := \frac{d}{dt} \int_{x \in V_t} \varphi \, dx = \int_{x \in V_t} \left(div(g \cdot \varphi) + \frac{\partial \varphi}{\partial t} \right) dx \quad (13a)$$

with $g = (g_1, \dots, g_n)$

$$\text{and} \quad div\,(g \cdot \varphi) = \sum_{i=1}^{n} \frac{\partial (g_i \varphi)}{\partial z_i}. \quad (13b)$$

It is near at hand to introduce the following special processes.

Definition 4. A continuous, deterministic Markov process $(\mathbb{R}, \mathbb{R}^n, F)$ is called φ-conservative, if $I_\varphi = 0$ for all $V_t \subset U_t$, otherwise dissipative $(I_\varphi < 0)$, accumulative $(I_\varphi > 0)$, respectively.

Note that in case $\varphi = 1$, (13a) is equivalent to the classical Liouville's theorem:

$$I_1 = \dot{V}_t = \int_{V_t} div\ g \cdot dx = 0 \quad (13c)$$

Because of $div\ grad^* H = 0$, we conclude readily the well-established fact: Any vector field g related to a process Ξ of classical mechanics is source-free and, consequently, such processes are 1-conservative.

If Markov process Ξ is conservative with respect to φ, then

$$\Phi_t := \int_{x \in V_t} \varphi \, dx \quad (14a)$$

is an integral invariant (of order n). By definition, with (13a) and (14a), Φ_t is an integral invariant of the process with the vector field g, if

$$\frac{\partial \varphi}{\partial t} = - div(g \cdot \varphi) \quad (14b)$$

in U_t. More specifically, if g is the vector field of a process of classical mechnics, i.e. $g = grad^* H$, then (14a) implies

$$\frac{\partial \varphi}{\partial t} = -[H, \varphi] \quad (15)$$

where $[.,.]$ denotes the Poisson bracket. Clearly, (15) is among others satisfied for any integral invariants of the given process.

Now, let us discuss (15) for stochastic Markov processes, which are continuous and deterministic in addition. Such processes are characterized (in simple cases) by a probability density of the form

$$f'(t_1,x_1;t_2,x_2):=\delta[F(t_2,t_1,x_1)-x_2]\cdot f(t_1,x_1) \qquad (16)$$

where δ denotes the delta-function, f the density of a stochastic Markov process, and F the transition operator of a continuous, deterministiv Markov process.

(16) implies

$$f(t_2,t_1)=\int_{\mathbb{R}^n}\delta(F(t_2,t_1,x_1)-x_2)\cdot f(t_1,x_1)\,dx_1 \qquad (17a)$$

and

$$\int_{U_{t_1}} f(t_1,x_1)dx_1=\int_{U_{t_2}} f(t_2,x_2)dx_2\,. \qquad (17b)$$

Clearly, the probability

$$P_t(U_t)=P\{\xi(t)\in U_t\}=\int_{U_t} f(t,x)dx \qquad (17c)$$

is an integral invariant, that is, (14b) and (15) are satisfied with $\varphi = f$.

Theorem 2. Let Ξ be a stochastic and continuous, deterministic Markov process with the Hamiltonian field g given by (12), and the density f(t,x). Then this density f satisfies

$$\frac{\partial f}{\partial t}=-[H,f]\,. \qquad (18b)$$

If (18b) is rewritten in the form

$$j\frac{\partial f}{\partial t}=L_H(f) \qquad (18c)$$

where $L_H = -j[H,.]$ denotes the Liouville's operator, then we obtain an equation which can be read as the Liouville's equation of statistical mechanics. L_H is a Hermitian operator if f vanishes on the boundary of U_t.

6. Quantum mechanics

According to Paragraph 5, a smooth functional $\varphi: T\times\mathbb{R}^n\rightarrow\mathbb{R}$ of a smooth and deterministic Markov process with the vector field g is an integral invariant, if (14b) is satisfied. However, in equation (14b), the operator div(g(.)) is not Hermitian. But it turns out that the following theorem can be proved:

Theorem 3. φ is a solution of

$$\frac{\partial\varphi}{\partial t} = -\mathrm{div}\,(g\cdot\varphi) \qquad (19)$$

if and only if $\varphi=|\psi|^2$ and ψ is a solution of

$$j\frac{\partial\psi}{\partial t}=L_g(\psi) \qquad (20a)$$

where the linear operator L_g is given by

$$L_g(\psi)=-jg\,\mathrm{grad}\,\psi-\frac{\psi}{2}(\mathrm{div}\,jg-V)\,. \qquad (20b)$$

In (20b), V = V(t,x) denotes an arbitrary real scalar field, and L_g is Hermitian if ψ vanishes on the boundary of its domain.

Proof: In view of (20a) and (20b), it is obvious that (19) holds with $\varphi=\psi\psi^*=|\psi|^2, \psi\in\mathbb{C}$. Conversely, (19) implies (20) with an arbitrary real function on t and x. Furthermore, the Hermitian property of L_g is established by a simple transformation of the integral $\int_U \psi^* L_g(\psi)dx$.

Thus we may claim: If Ψ is a solution of (20), then $\varphi = |\Psi|^2$ defines an integral invariant of the Markov Process Ξ, which is characterized by the vector field g.

Multiplying (20) with the Plank constant $\hbar$, we derive at

$$j\hbar \frac{\partial \Psi}{\partial t} = H_g(\Psi) \quad , \quad H_g = jL_g . \tag{21}$$

This equation can read as equation of quantum mechanics, where Hamiltonian H_g has, by virtue of (20b), in general the form

$$H_g(\Psi) = -jg\hbar \, \mathrm{grad}\, \Psi - \frac{\Psi}{2}(\mathrm{div}\, j\hbar g - V') \quad , \quad V' = \hbar V. \tag{22}$$

Clearly, any specification of g yields a special Hamiltonian H_g. We demonstrate this by a simple example.

Example a) Particle of mass m in a potential U (Schroedinger equation). By setting

$$g = \frac{\hbar}{m}\, \mathrm{grad}\, S \quad , \quad S = \mathrm{Im}\{\ln \Psi\} \tag{23}$$

(22) yields

$$H_g = -\frac{\hbar^2}{2m}\Delta + U' \tag{24}$$

where the scalar field U'can be chosen freely.

b) Particle of mass m in a potential U, a magnetic field described by the vector potential A. With

$$g = \frac{\hbar}{m}\, \mathrm{grad}\, S + \frac{2e}{c}A \quad , \quad \mathrm{div}\, A = 0 \quad , \quad S = \mathrm{Im}\{\ln \Psi\}$$

by virtue of (22), the Hamiltonian of this system is given by

$$H_g = -\frac{\hbar^2}{2m}\Delta - j\hbar\frac{2e}{c}A\, \mathrm{grad} + U' \tag{25a}$$

where again, U'can be chosen arbitrary. With

$$U' = \frac{e^2}{c^2}A^2 + U$$

one obtains the well-known Hamiltonian /3/:

$$H_g = \frac{1}{2m}\left(-j\hbar\, \mathrm{grad} - \frac{e}{c}A\right)^2 + U . \tag{25b}$$

References

/1/ Saizew, G.A.: Algebraische Probleme der mathematischen und theoretischen Physik, Akademie-Verlag, Berlin 1979

/2/ Lewin, A., Schwarz, W., Wunsch, G.: Stochastische Signale und Systeme in der Übertragungs- und Steuerungstechnik, Akademie-Verlag, Berlin (in Vorb.)

/3/ Wunsch, G.: Verallgemeinerung des Fundamentalsatzes von Kolmogoroff, Wiss. Z.d.TU Dresden, 38 (1989), 1, S.233-235

/4/ Wunsch, G.: Universelle Markov-Prozesse. In: Philippow: Contributions to the theory of electrical engineering, VEB Verlag Technik, Berlin (in Vorb.)

/5/ Hoare, C.A.R.: Communicating sequential processes, Prentice Hall International, London 1985

/6/ Arnold, V.I.: Gewöhnliche Differentialgleichungen, VEB Deutscher Verlag der Wissenschaften, Berlin 1979

/7/ Markov process, Liouville's theorem and dissipativ flows, Systems Analysis - Modelling - Simulation 6 (1989), 11

/8/ Landau, L.D.; Lifschitz, E.M.: Quantenmechanik, Akademie-Verlag, Berlin 1965

1 **Inverse and Improperly Posed Problems in Differential Equations**
Proceedings of the Conference on Mathematical and Numerical Methods, held in Halle/Saale (GDR) from May 29 to June 2, 1979
Edited by Gottfried Anger
296 pages, 1979

2 **Fundamentals of Computation Theorie FCT '79**
Proceedings of the Conference on Algebraic, Arithmetic and Categorial Methods in Computation Theory, held in Berlin/Wendisch-Rietz (GDR) September 17–21, 1979
Edited by Lothar Budach
576 pages, 1979

3 **Decidability and Generalized Quantifiers**
By Andreas Baudisch, Detlef Seese, Hans-Peter Tuschik, and Martin Weese
XII, 235 pages, 1980

4 **Zur Elastizitäts- und Thermoelastizitätstheorie**
Potentialtheoretische Methoden in Anwendung auf verschiedene Modelle
Von Lothar Jentsch und Johannes Maul
157 Seiten, 1980

5 **Systems Analysis and Simulation 1980**
Proceedings of the International Symposium held at Berlin (GDR) September 1–5, 1980
Edited by Achim Sydow
496 pages, 1980

6 **Nichtlineare diskrete Optimierung**
von Eberhard Girlich und Michail M. Kowaljow
218 Seiten, 1981

7 **Completeness Problems in the Structural Theory of Automata**
By Jürgen Dassow
148 pages, 1981

8 **Darstellung von Lösungen linearer elliptischer Differentialgleichungen**
Von Günther Wildenhain
92 Seiten, 1981

9 **Introduction to the Spectral Theory of Operators in Spaces with an Indefinite Metric**
By I. S. I[illegible]vidov, M. G. Krein, and H. Langer
120 pages, 1982

10 **Nonlinear Time-discrete Systems**
A General Approach by Nonlinear Superposition
By Michael Gössel
115 pages, 1982

11 **Preprimal Algebras**
By Klaus Denecke
162 pages, 1982

12 **Ergodic Theory and Related Topics**
Proceedings of the Conference held in Vitte/Hiddensee (GDR) October 19–23, 1981
Edited by Horst Michel
232 pages, 1982

13 **Singularly Perturbed Differential Equations**
By H. Goering, A. Felgenhauer, G. Lube, H.-G. Roos, and L. Tobiska
176 pages, 1983

14 **Leben wir in einer Volterra-Welt?**
Ein ökologischer Zugang zur angewandten Systemanalyse
Von Manfred Peschel und Werner Mende
233 Seiten, 1983

15 **Markov Decision Problems with Countable State Spaces**
Optimality Criteria – Algorithms – Clustering
By Hans Michael Dietz and Volker Nollau
174 pages, 1983

16 **Stochastische Verfahren zur Suche nach dem Minimum einer Funktion**
von Ryszard Zieliński und Peter Neumann
133 Seiten, 1983

17 **Invertibility and Asymptotics of Toeplitz Matrices**
By Albrecht Böttcher and Bernd Silbermann
200 pages, 1983

18 **Structural Induction on Partial Algebras**
Introduction to Theory and Application of Partial Algebras – Part II
By Horst Reichel
205 pages, 1984

19 **Algebraic Methods for Toeplitz-like Matrices and Operators**
By Georg Heinig and Karla Rost
212 pages, 1984

20 **Frege Conference 1984**
Proceedings of the International Conference held at Schwerin (GDR), September 10–14, 1984
Edited by Gerd Wechsung
408 pages, 1984

21 **Numerical Analysis of Selected Semilinear Differential Equations**
Edited by Christian Großmann, Thomas Riedrich, and Hartmut Schönheinz
203 pages, 1985

22 **Singular Ordinary Differential Operators and Pseudodifferential Equations**
By Johannes Elschner
200 pages, 1985

23 **Lotka-Volterra-Approach to Cooperation and Competition in Dynamic Systems**
Proceedings of the 5th Meeting of UNESCO's Working Group on Systems Theory held on the Wartburg, Eisenach (GDR) at March 5–9, 1984
Edited by Werner Ebeling and Manfred Peschel
271 pages, 1985

24 **Convergence Structures 1984**
Proceedings of the Conference on Convergence held in Bechyně (Czechoslovakia), September 24–28, 1984
Edited by Josef Novák, Werner Gähler, Horst Herrlich, and Jan Mikusiński
254 pages, 1985

25 **Parcella '84**
Proceedings of the II. International Workshop on Parallel Processing by Cellular Automata and Arrays held in Berlin (GDR), September 25–27, 1984
Edited by Wolfgang Händler, Tamás Legendi, and Gottfried Wolf
239 pages, 1985

26 **Multiobjective and Stochastic Optimization Based on Parametric Optimization**
By Jürgen Guddat, Francisco Guerra Vasquez, Klaus Tammer, and Klaus Wendler
175 pages, 1985

27/28 **Systems Analysis and Simulation 1985**
I: Theory and Foundations
II: Applications
Proceedings of the International Symposium held in Berlin (GDR), August 26–31, 1985
Edited by Achim Sydow, Manfred Thoma, and Robert Vichnevetsky
per volume 495 pages, 1985

29 **Parcella '86**
Proceedings of the III. International Workshop on Parallel Processing by Cellular Automata and Arrays held in Berlin (GDR), September 9–11, 1986
Edited by Tamás Legendi, Dennis Parkinson, Roland Vollmar, and Gottfried Wolf
279 pages, 1986 ISBN 3-05-500175-3

30 **Fuzzy Sets Applications, Methodological Approaches, and Results**
Proceedings of the International Workshop on Fuzzy Sets Applications cosponsored by IIASA and convened in cooperation with IAEA held on the Wartburg, Eisenach (GDR), at March 3–8, 1985
Edited by Steffen Bocklisch, Sergei Orlovski, Manfred Peschel, and Yasushi Nishiwaki
233 pages, 1986 ISBN 3-05-500016-16-1

31 **Mathematical Methods of Specification and Synthesis of Software Systems '85**
Proceedings of the International Spring School, held in Wendisch-Rietz (GDR), April 22–26, 1985
Edited by Wolfgang Bibel and Klaus P. Jantke
245 pages, 1986 ISBN 3-05-500018-8

32 **A Model Theoretic Oriented Approach to Partial Algebras**
Introduction to Theory and Application of Partial Algebras – Part I
By Peter Burmeister
319 pages, 1986 ISBN 3-05-500176-1

33 **Finite Difference Method on Irregular Networks**
A Generalized Approach to Elliptic Problems of Second Order
By Bernd Heinrich
206 pages, 1986 ISBN 3-05-500328-4

34 **Modern Algorithms for Large Sparse Eigenvalue Problems**
By Arnd Meyer
125 pages, 1987 ISBN 3-05-500332-2

35 **Parametric Optimization and Related Topics**
Edited by Jürgen Guddat, Hubertus Th. Jongen, Bernd Kummer, and František Nožička
407 pages, 1987 ISBN 3-05-500337-3

36 **Discretization in Differential Equations and Enclosures**
Edited by Ernst Adams, Rainer Ansorge, Christian Großmann, and Hans-Görg Roos
250 pages, 1987 ISBN 3-05-500362-4

37 **Factorization of Measurable Matrix Functions**
By Georgi Semenovich Litvinchuk and Ilya Matveyevich Spitkovskiĭ
372 pages, 1987 ISBN 3-05-500368-3

38 **Parallel Algorithms and Architectures**
Proceedings of the International Workshop on Parallel Algorithms and Architectures held in Suhl (GDR), May 25–30, 1987
Edited by Andreas Albrecht, Hermann Jung, and Kurt Mehlhorn
205 pages, 1987 ISBN 3-05-500397-7

39 **Parametric Integer Optimization**
By Bernd Bank and Reinhard Mandel
138 pages, 1988 ISBN 3-05-500398-5

40 **Computer Analysis of Images and Patterns 1987**
Proceedings of the II. International Conference CAIP '87 on Automatic Image Processing, held in Wismar (GDR), September 2–4, 1987
Edited by Leonid P. Jaroslavskiĭ, Azriel Rosenfeld, and Wolfgang Wilhelmi
282 pages, 1987 ISBN 3-05-500451-5

41 **Multivariable Control**
A Graph-theoretic Approach
By Kurt J. Reinschke
274 pages, 1988 ISBN 3-05-500452-3

42 **Equilibrium Distributions of Branching Processes**
By Alfred Liemant, Klaus Matthes, and Anton Wakolbinger
240 pages, 1988 ISBN 3-05-500453-1

43 **Andreotti-Grauert Theory by Integral Formulas**
By Gennadi Markovič Henkin and Jürgen Leiterer
270 pages, 1988 ISBN 3-05-500510-4

44 **Continual Means and Boundary Value Problems in Function Spaces**
By Efim Mikhailovich Polishchuk
164 pages, 1988 ISBN 3-05-500512-0

45 **Advances in Mathematical Optimization**
Invited papers dedicated to Prof. Dr. Dr. h. c. F. Nožička on occasion of his 70th birthday
Edited by Jürgen Guddat et al.
332 pages, 1988 ISBN 3-05-500543-0

46/47 **System Analysis and Simulation 1988**
Proceedings of the International Symposium held in Berlin (GDR), September 12–16, 1988
I: Theory and Foundations
II: Applications
Edited by Achim Sydow, Spyros G. Tzsafestas, and Robert Vichnevetsky
426/425 pages, 1988 ISBN 3-05-500654-2
ISBN 3-05-500655-0

48 **Parcella '88**
Proceedings of the IV. International Workshop on Parallel Processing by Cellular Automata and Arrays, held in Berlin (GDR), October 17–21, 1988
Edited by Gottfried Wolf, Tamás Legendi, and Udo Schendel
379 pages, 1988 ISBN 3-05-500656-9

49 **Algebraic and Logic Programming**
Proceedings of an International Workshop held in Gaussig (GDR), November 14–18, 1988
Edited by Jan Grabowski, Pierre Lescanne, and Wolfgang Wechler
278 pages, 1988 ISBN 3-05-500655-8

50 **Asymptotics for Elliptic Mixed Boundary Problems**
Pseudo-Differential and Mellin Operators in Spaces with Conormal Singularity
By Stephan Rempel and Bert-Wolfgang Schulze
418 pages, 1989 ISBN 3-05-500676-3

51 **Geobild '89**
Proceedings of the 4th Workshop on Geometrical Problems of Image Processing held in Georgenthal (GDR), March 13–17, 1989
Edited by Albrecht Hübler, Werner Nagel, Brian David Ripley, and Günter Werner
198 pages, 1989 ISBN 3-05-500677-1

52 **Splines in Numerical Analysis**
Contributions to the International Seminar ISAM-89 held in Weissig (GDR), April 24–28, 1989
Edited by Jochen W. Schmidt and Helmut Späth
198 pages, 1989 ISBN 3-05-500679-8

53 **Complex Methods on Partial Differential Equations**
Aspects in Complex Analysis
Edited by Claudio I. Withalm
258 pages, 1989 ISBN 3-05-500680-1

54 **Markov Processes and Control Theory**
Edited by Heinz Langer and Volker Nollau
234 pages, 1989 ISBN 3-05-500754-9

55 **Computer Analysis of Images and Patterns**
Proceedings of the III. International Conference CAIP '89 on Automatic Image Processing held in Leipzig (GDR), September 8–10, 1989
Edited by Klaus Voss, Dmitry Chetverikov, and Gerald Sommer
259 pages, 1989 ISBN 3-05-500755-7

56 **Quaternionic Analysis and Elliptic Boundary Value Problems**
By Klaus Gürlebeck and Wolfgang Sprößig
254 pages, 1989 ISBN 3-05-500756-5

57 **Computers in Science and Higher Education**
Contributions to the Conference BIT '89 held in Berlin (GDR), June 1989
Edited by Jan Grabowski
321 pages, 1989 ISBN 3-05-500774-3

58 **Wissenschaftliches Rechnen mit Ergebnisverifikation**
herausgegeben von Ulrich Kulisch
VI, 262 Seiten, 1989 ISBN 3-05-500816-2

The present collection of papers resulted from a seminar on nonlinear dynamics and quantum dynamical systems.
The papers on nonlinear dynamics mainly deal with the following subjects: generalized Poincaré criterion and strange attractor, Duffing equation and Hausdorff dimension, numerical methods for heteroclinic orbits, dynamical systems and integro-differential equations with cylindric phase space, bifurcation problems and invariant manifolds, integral invariants, system theory.
The papers on quantum dynamical systems contain contributions to the following problems:
the concept of entropy of operator algebras, properties of positive and completely positive maps in topological algebras, representation of automorphism groups and derivations, structure of state spaces.
The collection contains research papers as well as review articles.

Der vorliegende Sammelband enthält Arbeiten eines Seminars zur nichtlinearen Dynamik und quantendynamischen Systemen.
Die Arbeiten zur nichtlinearen Dynamik behandeln im wesentlichen folgende Aspekte:
verallgemeinertes Poincaré-Kriterium und seltsamer Attraktor, Duffing-Gleichung und Hausdorff-Dimension, numerische Methoden für heterokline Orbits, dynamische Systeme und Integro-Differentialgleichungen mit zylindrischem Phasenraum, Bifurkationsprobleme und invariante Mannigfaltigkeiten, Integralinvarianten, Systemtheorie.
Die Arbeiten zu quantendynamischen Systemen enthalten Beiträge zu folgenden Problemen: das Entropiekonzept für Operatoralgebren, Eigenschaften positiver und vollständig positiver Abbildungen in topologischen Algebren, Darstellung von Automorphismengruppen und Derivationen, Struktur von Zustandsräumen.
Der Sammelband enthält Beiträge sowohl zu Einzelaspekten der Theorie als auch Artikel mit Übersichtscharakter.

ISBN 3-05-500858-8
ISSN 0138-3019